BIOLOGICAL CONTROL OF POSTHARVEST DISEASES
THEORY AND PRACTICE

Edited by

Charles L. Wilson, Ph.D.
Michael E. Wisniewski, Ph.D.
Agricultural Research Service
U.S. Department of Agriculture
Appalachian Fruit Research Station
Kearneysville, West Virginia

CRC Press
Boca Raton Ann Arbor London Tokyo

Library of Congress Cataloging-in-Publication

Biological control of postharvest diseases – theory and practice/
edited by Charles L. Wilson and Michael E. Wisniewski.
    p. cm.
    Includes bibliographical references and index.
    ISBN 0-8493-4567-7
    1. Fruit—Postharvest diseases and injuries—Biological control.
    2. Vegetables—Postharvest diseases and injuries—Biological
    control. I. Wilson, Charles L. II. Wisniewski, Michael E.
    SB608.F8B56 1994
    634.046—dc20   93-33268
                                                                 CIP

© 1994 by CRC Press, Inc.

No claim to original U.S. Government works
International Standard Book Number 0-8493-4567-7
Library of Congress Card Number 93-33268
Printed in the United States of America 1 2 3 4 5 6 7 8 9 0
Printed on acid-free paper

PREFACE

Research in the past has concentrated on the production of food, with disproportionately less attention given to its preservation. Yet, it is estimated that 25% of our harvested fruits and vegetables are lost to postharvest spoilage in the U.S. and over 50% in developing countries. Food surpluses in more developed countries have concealed the magnitude of postharvest losses and diminished concern over combating them. However, recently a number of events have focused attention on postharvest losses of our fruits and vegetables.

Public concern over food safety has caused greater scrutiny of the pesticides used to preserve our food. A 1989 National Academy of Sciences (NAS) report focused attention on the oncological risk from pesticide residues in our food. Fungicides which had been perceived as relatively safe were targeted in this report as posing a greater oncological risk than both herbicides and insecticides together. Subsequent to this, a number of key fungicides used to control postharvest rots of fruits and vegetables were withdrawn from the market.

Biological control offers a logical alternative to synthetic fungicides for the control of postharvest diseases. Our pursuit of biological alternatives shouldn't be limited by our definition of biological control. Baker has defined biological control as "... the decrease of inoculum or the disease-producing activity of a pathogen accomplished through one or more organisms, including the host plant but excluding man." Utilizing this definition, a number of approaches are available for finding alternatives to synthetic fungicides. Among these are the use of (1) antagonistic microorganisms; (2) inherent or induced resistance; and (3) natural plant- or animal-derived fungicides.

Researchers in biological control can commonly be heard to say, "What we need are more success stories." Few biological control methods for plant diseases have made it from theory to practice. Inconsistent results when biological control methods are applied in the field have been attributed to the inherent variability of living systems and the vagaries of the environment. When biological control is pitted with chemical control it has not generally matched up in efficacy or consistency, there is reason to believe that prospects may be brighter in the biological control of postharvest diseases.

In their 1983 book, *Biological Control of Plant Pathogens,* Cook and Baker listed only one example of the use of an antagonistic microorganism to control a postharvest disease of fruits or vegetables. Since that time research in this field has exploded. You will find in this book numerous reports of the discovery and development of antagonistic fungi and bacteria for the control of postharvest diseases. Some of these organisms have been patented and tested under semi-commercial conditions and attempts are being made to commercialize them. As we move toward the application of this new technology, new research problems are emerging which must be solved. The perspective of researchers, industry, and regulatory agencies toward this emerging technology is presented in this treatise.

We can approach the use of antagonistic microorganisms to control postharvest diseases of fruits and vegetables with considerable optimism. Although products for the biological control of plant diseases have been slow in reaching the marketplace, there is reason to believe that postharvest biological control with antagonists may be on a faster track. Part of this is due to the fact that laboratory results on postharvest diseases are more easily translated into postharvest treatments than into field treatments.

Environmental conditions in the postharvest environment are easier to control and closer to those in the laboratory than in the field. Also, the postharvest environment is more conducive to the application of biological control agents. Target areas for the application of antagonists are more concentrated and exposed than those encountered for plants in the field. The higher value of harvested commodities also makes elaborate biological control procedures more economically feasible than they would be in the field.

Resistance in harvested commodities to postharvest pathogens is poorly understood and constitutes an underutilized means of controlling postharvest diseases. Most of our understanding of resistance in plants has come from studies of the vegetative plant body. We have only a rudimentary understanding of resistance in harvested commodities. In fact, in our breeding programs to select for less astringency and more tenderness in fruits and vegetables we may have robbed plants of phenolic compounds and thick cell walls which impart resistance.

The pioneering work of Drs. Stevens, Lu, and Khan at Tuskegee University has highlighted the potential of utilizing induced resistance in harvested commodities as a means of controlling postharvest diseases.These investigators and their associates have demonstrated that low-dose UV light can be utilized to induce resistance in postharvest commodities such as sweet potatoes, onions, citrus, apples, and peaches. The capability of UV light to induce resistance in such a wide range of commodities holds promise as an alternative to synthetic fungicides for the control of postharvest rots of fruits and vegetables.

Because of concerns over the safety of insecticides resulting from Rachel Carson's book *Silent Spring* in the 1960s, considerable effort was put into developing natural plant-derived insecticides. Because synthetic fungicides were perceived to be relatively safe at that time, no attempt was made to find safer alternatives. Recent concerns over the oncogenic risk posed by synthetic fungicides used to preserve our food has placed new emphasis on the need for safer plant-derived fungicides. Although plants are known to contain a number of very effective fungicides, none have been developed and made available commercially.

Recent successes in the biological control of postharvest disease of fruits and vegetables offer real alternatives to synthetic fungicides. A number of products are close to reaching the market and research in this area is accelerating. Biological control of postharvest diseases is bridging the gap from theory to practice. It is hoped that the momentum from these successes will carry us forward to products which provide safe and effective alternatives to synthetic fungicides for the control of postharvest diseases.

Although we can point to success in the biological control of postharvest diseases, we should not ignore our failures. Our inability to explain and control inconsistencies in biological control treatments often springs from a lack of fundamental understanding of the biology that underlies biological control. Once a crop is harvested, two forms of deterioration are accelerated. One is inherent in the commodity itself involving the senescence process and the other involves microbial deterioration. These two processes are inextricably bound since as harvested crops deteriorate through senescence their resistance to microbial invasion generally decreases. In order to intelligently pursue biological control procedures for postharvest diseases, fundamental knowledge is needed on the microecology of organisms associated with harvested crops and resistance to disease in harvested commodities as it relates to the senescing process.

This book is a compilation of the thinking of a wide array of experts in the areas of microecology, plant physiology, cytology, biochemistry, natural products, biocontrol, pesticide regulation, and postharvest pathology. Also, industry experts and regulatory personnel present their perspective's on the commercialization of biological control methods as alternatives to synthetic chemicals for the control of postharvest diseases of fruits and vegetables. It is hoped that it will serve as a springboard for students and researchers in this area for continued progress.

Charles L. Wilson
Michael E. Wisniewski

THE EDITORS

Charles L. Wilson, Ph.D., has been conducting research on the biological control of plant diseases and weeds for over 35 years. His early interest, as a Professor at the University of Arkansas (1958 to 1968), was in the biological control of weeds with plant pathogens. Dr. Wilson wrote the first comprehensive review on the biological control of weeds for *Annual Reviews* and organized the first international symposium on this subject. While at the University of Arkansas, Dr. Wilson received the highest award given by the Arkansas Alumni Association for distinguished research and teaching.

In 1968, Dr. Wilson joined the Agricultural Research Service of the U.S. Department of Agriculture as Investigations Leader for Shade Tree and Farm Windbreak Investigations. He also served as Director of the ARS Nursery Crops Laboratory in Delaware, Ohio. Dr. Wilson was responsible for a national program on shade tree improvement and nursery crop research. During this period, he conducted research on the biological control of dutch elm disease.

Dr. Wilson joined the USDA-ARS Appalachian Fruit Research Station in 1978 where he initiated a research program to find safe and effective alternatives to synthetic fungicides for the control of fruit diseases. This internationally recognized research program has yielded a variety of innovative approaches for the control of postharvest diseases of fruit which include the use of antagonistic microorganisms, natural plant-derived fungicides, and the use of induced resistance. Dr. Wilson received the Biological Sciences Award from the Washington Academy of Sciences in 1983 for innovative research in the manipulation of plant diseases. He was also made a Fellow of the Academy. In 1984, Dr. Wilson was named Scientist of the Year by the North Atlantic Area of the Agricultural Research Service for his innovative research in the biological control of postharvest diseases of fruit.

Dr. Wilson has written and lectured extensively on the biological control of postharvest diseases of fruits and vegetables. He has been cooperating with various scientists around the world interested in this subject including Canada, Israel, Italy, Egypt, Morocco, Australia, New Zealand, South Africa, Brazil, and Japan.

Michael Wisniewski, Ph.D., has been conducting research on biological control of postharvest diseases for over eight years. His principal effort has been in understanding the mode-of-action of antagonists that have been identified as highly successful biocontrol agents. In addition to his biocontrol research, Dr. Wisniewski has also conducted a successful and innovative program of research on cold acclimation in fruit trees.

Dr. Wisniewski joined the USDA-ARS, Appalachian Fruit Research Station in 1983. During his tenure he has developed a strong program of cold hardiness and biocontrol research. His collaborative research with Dr. Charles Wilson has been recognized internationally and he has lectured widely in the United States and Europe on the subject of postharvest biological control. Dr. Wisniewski has received several special awards by the USDA-ARS for his outstanding scientific accomplishments. Among these, he received in 1992, the USDA-ARS Early Career Scientist Award for innovative research on cold hardiness and biological control in fruit crops.

CONTRIBUTORS

Joseph Arul, Ph.D.
Department of Food Science and
Technology
Horticultural Research Centre
Laval University
Sainte-Foy, Quebec, Canada

David Bays, Ph.D.
Office of Pesticide Programs
U.S. Environmental Protection Agency
Washington, D.C.

Edo Chalutz, Ph.D.
Institute for Technology and Storage of
Agricultural Products
Agricultural Research Organization
The Volcani Center
Bet Dagan, Israel

Horace G. Cutler, Ph.D.
Agricultural Research Service
U.S. Department of Agriculture
Athens, Georgia

Richard A. Daoust, Ph.D.
Ecogen Inc.
Langhorne, Pennsylvania

Ernest Delfosse, Ph.D.
National Biological Control Institute
Animal, Plant, and Health Inspection
Service
U.S. Department of Agriculture
Hyattsville, Maryland

Samir Droby, Ph.D.
Institute for Technology and Storage of
Agricultural Products
Agricultural Research Organization
The Volcani Center
Bet Dagan, Israel

Bertold Fridlender, Ph.D.
Ecogen Israel Partnership
Jerusalem, Israel

Ahmed El Ghaouth, Ph.D.
Department of Food Science and
Technology
Horticultural Research Centre
Laval University
Sainte-Foy, Quebec, Canada

Richard J. Gouger, Ph.D.
Ecogen Inc.
Langhorne, Pennsylvania

Carl Grable, Ph.D.
Office of Pesticide Programs
U.S. Environmental Protection Agency
Washington, D.C.

Janelle S. Graeter, M.S.
Office of Cooperative Interactions
Agricultural Research Service
U.S. Department of Agriculture
Beltsville, Maryland

Robert A. Hill, Ph.D.
The Horticulture and Food
Research Institute of
New Zealand Ltd.
Hamilton, New Zealand

Raphael Hofstein, Ph.D.
Ecogen Israel Partnership
Jerusalem, Israel

Phillip Hutton, M.S.
Office of Pesticide Programs
U.S. Environmental Protection Agency
Washington, D.C.

John Kough, Ph.D.
Office of Pesticide Programs
U.S. Environmental Protection Agency
Washington, D.C.

Michael Mendelsohn, B.S.
Office of Pesticide Programs
U.S. Environmental Protection Agency
Washington, D.C.

Harold E. Moline, Ph.D.
Horticultural Crops Quality
Laboratory
Agricultural Research Service
U.S. Department of Agriculture
Beltsville, Maryland

Gail E. Poulos, B.A.
Office of Cooperative Interactions
Agricultural Research Service
U.S. Department of Agriculture
Beltsville, Maryland

P. Lawrence Pusey, Ph.D.
Tree Fruit Research Laboratory
Agricultural Research Services
U.S. Department of Agriculture
Wenatchee, Washington

Peter G. Sanderson, Ph.D.
Mid-Columbia Agricultural Research
  and Extension Center
Oregon State University
Hood River, Oregon

Joseph L. Smilanick, Ph.D.
Horticultural Crops Research
  Laboratory
Agricultural Research Service
U.S. Department of Agriculture
Fresno, California

Robert A. Spotts, Ph.D.
Mid-Columbia Agricultural Research
  and Extension Center
Oregon State University
Hood River, Oregon

Harvey W. Spurr, Jr., Ph.D.
Department of Plant Pathology
North Carolina State University
  and
Crops Research Laboratory
Agricultural Research Service
U.S. Department of Agriculture
Oxford, North Carolina

Steven K. Whitesides, Ph.D.
Ecogen Inc.
Langhorne, Pennsylvania

CONTENTS

Chapter 1

Emerging Technologies for the Control of Postharvest Diseases of Fresh Fruits and Vegetables

Joseph Arul

CONTENTS

I. INTRODUCTION

Fruits and vegetables are an important part of the human diet. They add flavor and variety and supply essential nutrients such as vitamins and minerals. They are also a major source of complex carbohydrates, antioxidants, and anticarcinogenic substances which are important to human health and well-being. Increasing awareness by consumers that diet and health are linked has resulted in greater consumption of fruits and vegetables. Fresh produce is gaining in consumption over processed products, as the consumer prefers the wholesomeness of fresh produce. The per capita consumption of fresh vegetables in the U.S. increased from 1978 to 1988 by 26% while canned vegetable consumption declined by about 5% in the same decade.[1] At the same time, consumers are also concerned about the safety of the foods they eat. This concern is an important issue to the fresh produce sector, more than any other segment of the food industry. The public demands foods free of microbial toxins, pathogens, and chemical residues. The "Fresh Trends '90" survey found a large majority of its respondents concerned about chemical residues on their produce.[3] To meet the growing demand for fresh produce free from chemical residues, strategies that focus solely on increased crop production may not be adequate if postharvest preservation is not taken into consideration. Safer and more efficient preservation techniques must be developed.

Fruits and vegetables are highly perishable and maintain an active metabolism in the postharvest phase. The major factors causing the early termination of the storage life of fresh produce are fungal infection, senescence, and transpiration. Storage diseases, especially those caused by fungal pathogens, are responsible for substantial postharvest losses. Factors that accelerate senescence and favor microbial growth, such as physiological and mechanical injuries, as well as exposure to undesirable storage conditions (high ambient temperature and high humidity), can promote postharvest decay. Postharvest losses are significant in terms of both energy input and time. Postharvest deterioration is not only a problem for the producers, but also it persists throughout the distribution chain affecting the cost and availability of the produce to the consumer and the ability of producers to service distant markets.

1

The horticultural industry is undergoing new developments in mechanical harvesting and bulk handling. Unfortunately, these developments can aggravate postharvest diseases because of increased wounding and creation of environments favorable for disease development. Thus, our capability to control diseases could ultimately be a key factor dictating the way horticultural crops are handled and the adoption of new developments in handling, storage, and transportation methodology by industry. Mechanical harvesting, as practiced today, is responsible for a high incidence of wounding to crops. This necessitates prompt refrigeration and treatment with fungicide to alleviate potential losses due to decay. In the context of public resistance to chemical residues in food and environmental safety, there is a critical need to develop alternative methods to control diseases. The need becomes urgent when one considers the possible deregistration of effective fungicides such as benomyl and the development of fungicide-tolerant strains of postharvest pathogens.[4]

This chapter presents an overview of some of the novel approaches emerging as possible alternatives to synthetic fungicides. They include intensification of the natural defense mechanisms of harvested produce, use of natural biocides, genetic transformation, and the use of biological antagonists. Subsequent chapters in this book discuss the scientific basis and potential application of this emerging technology for controlling postharvest diseases.

II. INTENSIFICATION OF THE NATURAL DEFENSE MECHANISMS OF THE HOST

Although harvested crops possess several mechanisms to defend themselves against invasion by pathogens, this potential has not been fully appreciated by postharvest pathologists. Most of our knowledge of plant resistance mechanisms has been gained from studies of the vegetative plant body where considerable attention has been placed on induced resistance as a form of crop protection. Plants have acquired a variety of physical and biochemical strategies to defend themselves against microbial attack.[5] These include the accumulation of phytoalexins, modification of host cell wall, and synthesis of antifungal hydrolases such as chitinases and β-1,3-glucanases. The intrinsic and variable nature of defense mechanisms and their regulation in harvested crops remain sketchy at this point.

The expression of resistance in harvested crops appears to be intimately linked to the ripening process. With few exceptions, the susceptibility of fruit and vegetables to postharvest decay dramatically increases with increased senescence of the tissue.[6] This is believed to be associated with a decreased ability of tissue to synthesize active and preformed inhibitors. Preformed inhibitors provide disease resistance to young tissue but degrade with age.[7-9] The degradation of preformed inhibitors appears to account for the resumption of latent infections in many crops.[6] Thus maintaining constitutive and inducible defense responses of the host by slowing down senescence could be of significance in reducing postharvest decay. This can be accomplished by manipulating storage conditions such as temperature and atmosphere.[10-13] Although the beneficial effects of these practices in reducing decay losses have been demonstrated with several crops, they are not sufficient to protect harvested tissue from infection and in some cases they can enhance decay. For instance, the use of moisture impermeable plastic films for packaging can lead to serious problems of decay and mold growth as a result of condensation.[14] Pusey et al.[15] have recently reviewed our present technology for the preservation of harvested fruits and vegetables. They suggest that we reexamine established procedures for the preservation of harvested commodities, but they also recognize that new methodologies are needed.

The induction of resistance responses in harvested commodities holds great promise as a means of controlling postharvest diseases. Deliberate stimulation of antifungal enzymes and accumulation of phytoalexins to inhibitory level through prestorage treatment with defense response elicitors could give harvested tissue an advantage in restricting infection. Success with the induction of resistance in a number of harvested commodities with low-dose UV-C light points to the potential of this technology. Induced accumulation of phytoalexins with fungal cell wall hydrolysate in pepper fruit[16] and mango,[17] and by UV light in kumquat and orange fruit[18] and carrot[19-21] was found to increase disease resistance. However, the responsiveness of the postharvest tissue to UV-C treatments decreased with its physiological age.[21,22] Prestorage treatment with low-dose UV light also controlled natural infection in Walla Walla onions,[23] in sweet potatoes,[24] and in pepper and strawberry fruits.[25] Although the physiological basis of low-dose UV light action in the control of rots has not been elucidated, it may in part be a consequence of delay in the ripening process.[25-27]

Prestorage heating also shows potential as a nonchemical method for protecting the crops against both physiological and pathological disorders, intensifying natural resistance to infection, and slowing down the ripening process.[28] Heat treatment may be effective either by inhibiting directly the growth of pathogens,[29] or by inducing natural resistance of the fruit.[30,31] In spite of interesting possibilities with prestorage heating, the energy requirements of heat treatment and the refrigeration load to subsequently cool down the produce to storage temperatures may prove to be a liability.

In recent years, there has been a renewed interest in exploiting induced systemic resistance as a method for controlling plant diseases. Induced systemic resistance implies structural or chemical changes that occur after the plant is challenged by a potential pathogen.[32-34] Systemic responses develop in cells distant from the site of the stimulus and involve activation of chitinase and glucanase enzymes, peroxidases, and protease inhibitors. The possibility of inducing systemic disease resistance was shown recently in stored carrots inoculated with *Botrytis cinerea*.[35] Although it is inconceivable to preinoculate postharvest crops with pathogens, it would still be possible to treat the crops with nonpathogens, active elicitors, or antagonists which are capable of inducing systemic resistance. Eventually it may also be possible to treat crops with signal compounds involved in the systemic resistance response, once more information becomes available on signals in plant tissue.

As we learn more about elicitors and signal compounds involved in the induction of various disease resistance responses, the way signals are transmitted, the molecular events associated with the resistance responses, the contribution of various mechanisms to disease control in harvested crops, and the effect senescence or the aging process have on the disease resistance potential of the harvested crop, practical methods of controlling diseases will ultimately emerge.

III. NATURAL BIOCIDES

Plants produce a number of secondary metabolites which are biocidal. Detailed study of some of these compounds, particularly nitrogen-containing aromatics such as benzoxazinones, is believed to have contributed to the development of successful fungicides such as benomyl.[36,37] Effective insecticides have been derived from plants, but little effort has been made to develop plant-derived fungicides for practical use.[38] The existence of such compounds in a wide variety of plants has been clearly demonstrated, and they could prove useful alternatives to synthetic fungicides. Recently, hinokitiol, an antifungal compound derived from cedar, was shown effective in controlling postharvest decay.[39] A number of essential oils and volatiles such as acetaldehyde produced by plants possess fungicidal activity and may have a potential for the control of postharvest diseases.[40-43] Ultimately, their application will depend on their phytotoxicity and their risk to human health.

ʳFurthermore, the search for natural biocides should not be limited to plant sources. For instance, chitosan (an animal-derived polymer) has shown potential as a preservative by being both fungicidal and an elicitor of resistance responses in the host.[44] Again, the search should not be limited to natural compounds which are inherently fungitoxic, but should encompass compounds which can directly control diseases either by acting against the disease process or by activating resistance responses in the host.[45] For rapid development in this area, it is imperative that we devise effective bioassays designed to reveal the potential of natural compounds for direct or indirect control of diseases, and establish structure-activity relationships which could help to identify promising compounds or determine minor structural modifications of the natural product that may lead to enhanced activity.

A debate is in progress on the comparative safety of synthetic and natural pesticides. Ames and associates[46] contend that natural plant-derived pesticides are potentially as carcinogenic as synthetic pesticides. Others contend that because of coevolution, animals are better able to degrade natural pesticides. Also, natural pesticides are considered more readily degradable in the environment by microorganisms and photodecomposition. More research is needed to compare the safety and biodegradability of natural and synthetic pesticides. Nevertheless, the search for effective fungicides in plants should not be deterred by the debate over the relative safety of natural compounds. With the wide array of fungicidal compounds that have been found in plants, some undoubtedly will prove effective and safe for human consumption.

Recently, there has been a renewed interest in the use of edible coatings as alternatives to plastic film wrappings which are nonbiodegradable and contribute to the problem of solid waste disposal.[47] Coatings modify the internal atmosphere of the tissue and reduce water loss in harvested fruits and vegetables. However, most of the commercially used oil-based coatings often increase the incidence of decay.[48] El Ghaouth et al.[49-51] found that coating produce with chitosan delayed senescence, reduced weight loss, and controlled postharvest decay without causing any phytotoxicity or altering the ripening capacity of the fruit. Such a coating not only can be used to mimic the benefits of modified atmosphere, but also can serve as a fungicide.[52-55] Edible coatings could also serve as carriers for effective antagonistic microorganisms and natural biocides to ensure an even distribution of these agents over the fruit surface. Several antagonistic yeasts were found to be compatible with water-based coatings.[55a] Thus development of bioactive coatings could be another interesting approach to control decay and ripening.

IV. GENETIC TRANSFORMATION OF THE HOST

Genetic transformation of plants to increase disease resistance and to modify their ripening characteristics has received much interest in the last 5 years. The use of antisense ribonucleic acid (RNA) techniques has generated considerable information regarding the function and regulation of a number of ripening genes.[56,57] Transgenic tomato plants have been developed[58,59] that synthesize antisense RNA which downregulates the expression of polygalacturonase or 1-aminocyclopropane-1-carboxylate (ACC) synthase, the enzyme catalyzing the conversion of ACC to ethylene. Although antisense gene technology offers benefits of delayed ripening and senescence and thereby extends the shelf life of harvested crops, it is too early to embrace this approach designed solely to inhibit specific gene expressions. Inhibition of specific genes may affect desirable biochemical changes associated with fruit ripening, particularly flavor and texture development. Key processes regulating ripening and senescence, and the factors controlling the rate of molecular events accompanying senescence are yet to be identified.

If major gains in prolonging the storability of crops without compromising the desirable attributes of ripening are to be achieved, a better understanding of the factors involved in the initiation and regulation of postharvest senescence is required. The recent investigation by King and associates[60,61] in harvested asparagus spears is worthy of note. They observed that the mere harvesting process led to expression of mRNAs that were not observed in normal development and that harvest may code for enzymes involved in the subsequent deterioration processes. Since major changes in respiratory metabolism (which is closely linked to postharvest deterioration) and gene expression have been identified within hours after harvest, it may be possible in the future to identify factors controlling the rate of events associated with senescence, if not the sequence of events.[61] The rate of postharvest senescence can be inhibited either through genetic transformation or by traditional breeding.

Recent advances in molecular biology have also resulted in considerable progress in our understanding of the regulation of expression of genes involved in defense reactions.[62,63] The identity of these genes has been determined in a number of cases, and transgenic plants capable of constitutive expression of defense-related gene products such as chitinase have been developed.[64] The insertion of a chitinase gene in tobacco and canola plants was shown to result in enhanced resistance to *Rhizoctonia solani* but not to *Cercospora nicotinae*.[64] This suggests that other factors besides chitinase may be involved in disease resistance.

Because of the multicomponent nature of disease resistance, strategies that consist of introducing a single defense-suspected gene may not lead to the development of disease-resistant plants. More refined studies are needed to determine the mechanisms by which expression of defense genes is regulated and coordinated, as well as the role they play in resistance. The combination of site-directed mutagenesis and transformation of the host could help determine the role of any suspected defense gene in resistance. With recent advances in molecular biology, there is good hope for the development of disease-resistant plants which constitutively express defense genes in both vegetative and harvested plant parts. Since harvested commodities become more susceptible to infection as they ripen, one can speculate on the use of developmentally regulated promoters to drive the expression of defense genes. This could confer resistance to the tissue at a stage when it is most vulnerable to fungal attack. Such an approach will be possible once organ-specific and developmentally regulated promoters are readily available.

V. USE OF ANTAGONISTIC MICROORGANISMS

The use of naturally occurring antagonistic microorganisms to protect crops against disease is becoming a potentially new form of crop protection. In postharvest, a number of storage diseases have been controlled by introducing biological antagonists.[65-69] Several antagonistic microorganisms have been patented and are targeted for commercialization for control of postharvest diseases. However, most of the identified antagonists against postharvest pathogens have been tested only against a limited number of pathogens on certain crops.[70] Further research is needed to identify antagonists with a broad spectrum of activity against a wide range of pathogens on a variety of crops. The antagonistic action can be expressed in a number of ways.[71] Some antagonists such as *Bacillus subtilis* reduce infection by producing antibiotics,[72] while most involve nutrient competition and direct parasitism.[73,74] Evidence also exists that antagonistic yeasts may induce resistance responses in the host.[74a]

Any biological control procedure which is developed as an alternative will be compared with synthetic fungicides in terms of its effectiveness and reliability. The effectiveness of antagonists may be affected by storage conditions. Since low temperatures and in some cases, controlled atmospheres are emphasized for prolonged storage, it is important that the effectiveness of antagonists be tested under prevalent storage conditions rather than at ambient temperatures. For the antagonistic microorganisms to compete on these grounds, considerable research will be required to enhance their activity and to develop application procedures that produce effective and reliable control of diseases.

Subsequent chapters will discuss the considerable effort that is underway to develop antagonistic microorganisms as biological control agents for postharvest diseases. One approach consists of enhancing the antagonistic activity of antagonists by genetic transformation. This idea is addressed by Pusey (Chapter 7). In the development of this new technology, a number of critical issues need be addressed. How will the application of these "living fungicides" be regulated? How will the public react to this new technology? Since microorganisms have been traditionally used for preserving foods since ancient times, it is hoped that this technology will be understood and accepted by the public. These issues will be considered elsewhere in this book.

In order to sustain progress in biological control with antagonistic microorganisms, basic studies on microbial ecology which underlies this technology are required. An understanding of the microecology of plant surfaces is critical to the intelligent selection and application of antagonistic microorganisms. Spurr in Chapter 2 discusses the scientific basis of this approach which can foster further developments in this area.

VI. CONCLUSION

With impending restrictions being placed on the use of synthetic fungicides and pesticides, an intense effort is needed to find safer alternatives. A range of approaches is being actively pursued to achieve practical control of storage diseases. They include the use of cultivars with genetic resistance to diseases, activation of intrinsic defense mechanisms of the tissue, genetic manipulation, and use of natural biocides and antagonists. Since the onset of senescence is accompanied by an increase in susceptibility to infection, techniques that delay senescence can be particularly effective in controlling diseases. However, a combination of complementary techniques could lead to effective control of diseases. In addition, careful cultural and management practices such as harvesting at optimum maturity, precise control of storage temperatures, humidity and atmosphere, and disease avoidance by proper sanitation procedures can all be beneficial in reducing postharvest disease losses.

REFERENCES

1. Economic Research Service, U.S. Department of Agriculture, Vegetables and Specialities Situation and Outlook Yearbook, August 1990.
2. Economic Research Service, U.S. Department of Agriculture, Fruit and Tree Nuts Situation and Outlook Yearbook, August, 1989.
3. Vance Research Services and Market Facts, Inc., Fresh Trends '90: A Profile of the Fresh Produce Consumer, Reports 1–4, 1990.
4. Spotts, R.A. and Cervantes, L.A., Populations, pathogenecity, and benomyl resistance of Botrytis spp., Penicillium spp., and Mucor pirformis in packinghouses, Plant Dis., 70, 106, 1986.
5. Bailey, J.A. and Deverall, B.J., Eds., The Dynamics of Host Defense, Academic Press, New York, 1983.
6. Eckert, J.W. and Ratnayake, M., Host-pathogen interactions in postharvest diseases, in Post-Harvest Physiology and Crop Preservation, Lieberman, M., Ed., Plenum Press, New York, 1983, 247.
7. Davies, W.P., Infection of carrot roots in cool storage by Centrospora acerinia, Ann. Appl. Biol., 85, 163, 1977.
8. Dennis, C., Susceptibility of stored crops to microbial infection, Ann. Appl. Biol., 85, 430, 1977.
9. Goodliffe, J.P. and Heale, J.B., Factors affecting the resistance of cold-stored carrots to Botrytis cinerea, Ann. Appl. Biol., 87, 17, 1977.

10. **Burton, W.G.,** Some biophysical principles underlying the controlled atmosphere storage of plant material, *Ann. Appl. Biol.,* 78, 149, 1974.

11. **Kader, A.A.,** Biochemical and physiological basis for effects of controlled and modified atmospheres on fruits and vegetables, *Food Technol.,* 40, 99, 1986.

12. **Rhodes, M.J.C.,** The physiological basis for the conservation of food crops, *Prog. Food Nutr. Sci.,* 4, 11, 1980.

13. **El-Goorani, M.A. and Sommer, N.F.,** Effects of modified atmospheres on postharvest pathogens of fruits and vegetables, *Hortic. Rev.,* 3, 412, 1981.

14. **Marcellin, P.,** Conservation des fruits et legumes en atmosphère contrôlée â l'aide de membranes de polymères, *Rev. Gen. Froid,* 3, 217, 1974.

15. **Pusey, P.L., Wilson, C.L., and Wisniewski, M.E.,** Management of postharvest diseases of fruits and vegetables: strategies to replace vanishing fungicides, in *Pesticide-Plant Pathogen Interaction in Crop Production: Beneficial and Deleterious Effects,* CRC Press, Boca Raton, FL, (in press).

16. **Adikaram, N.K.B., Brown, A.E., and Swineburne, T.R.,** Phytoalexin induction as a factor in the protection *Capsicum annuum* L. fruits against infection by *Botrytis cinerea* Pers., *J. Phytopathol.,* 122, 267, 1988.

17. **Boulet, M., Arul, J., Verret, P., and Kane, O.,** Induced resistance of stored mango (*Magnifera indica* L.) fruits to mold infection by treatment with *Colletotrichum gloeosporioides* L. cell wall hydrolysate, *Can. Inst. Food Sci. Technol. J.,* 22, 161, 1989.

18. **Rodov, V., Ben-Yohoshua, S., Kim, J.J., Shapiro, B., and Ittah, Y.,** Ultraviolet illumination induces scoparone production in Kumquat and orange fruit and improves decay resistance, *J. Amer. Soc. Hort. Sci.,* 117, 788, 1992.

19. **Mercier, J., Ponnampalam, R., and Arul, J.,** Induction of phytoalexin production and disease resistance in carrot slices by UV light, *Proc. Am. Can. Phytopathol. Soc.,* Abstr., East Lansing, MI, 1990, C18.

20. **Mercier, J., Arul, J., Ponnampalam, R., and Boulet, M.,** Induction of 6-methoxymellein and resistance to storage pathogens in carrot slices by UV-C, *J. Phytopathol.,* 137, 44, 1993.

21. **Mercier, J., Arul, J., and Julien, C.,** Effect of UV-C on phytoalexin accumulation and resistance to *Botrytis cinerea* in stored carrots, *J. Phytopathol.,* 137, in press.

22. **Creasy, L.L. and Coffee, M.,** Phytoalexin production potential of grape berries, *J. Am. Soc. Hortic. Sci.,* 113, 230, 1988.

23. **Lu, J.Y., Stevens, C., Yakubu, P. and Loretan, P.A.,** Gamma, electron bean and ultraviolet radiation on control of storage rots and quality of Walla Walla onions, *J. Food Process. Preserv.,* 12, 53, 1987.

24. **Stevens, C., Khan, V.A., Tang, A.Y., and Lu, J.Y.,** The effect of ultraviolet radiation on mold rots and nutrients of stored sweet potatoes, *J. Food Prot.,* 53, 223, 1990.

25. **Arul, J., Mercier, J., Baka, M., and Maharaj, R.,** Photochemical therapy in the preservation of fresh fruits and vegetables: disease resistance and delayed senescence, *Proc. Int. Symp. Physiological Basis of Postharvest Technologies (Postharvest 1992),* Abstr., University of California, Davis, CA, 1992, 42.

26. **Lu, J.Y., Stevens, C., Khan, V.A., Kabwe, M., and Wilson, C.L.,** The effect of ultraviolet irradiation on shelf-life and ripening of peaches and apples, *J. Food Qual.,* 14, 229, 1991.

27. **Maharaj, R., Arul, J., and Nadeau, P.,** Photochemical therapy in the preservation of fresh tomatoes by delaying senescence, *IFT Annu. Meet.,* Chicago, July 10–14, 1993.

28. **Klein, J.D. and Lurie, S.,** Heat treatments for improved postharvest quality and horticultural crops, *HortTechnology,* 2, 316, 1992.

8

29. **Couey, H.M.,** Heat treatment for control of postharvest diseases and insect pests of fruits, *HortScience,* 24, 198, 1989.
30. **Ben-Yohoshua, S., Shapiro, B., Kim, J.J., Sharoni, J., Carmeli, S., and Kashman, Y.,** Resistance of citrus fruit to pathogens and its enhancement by curing, in *Proc. 6th Int. Citrus Congr.,* Goren, R. and Mendel, K., Eds., Balaban Publ., Philadelphia, 1988, 1371.
31. **Spotts, R.A. and Chen, P.M.,** Prestorage heat treatments for control of decay of pear fruit, *Phytopathology,* 77, 1578, 1987.
32. **Hammerschmidt, R. and Stermer, B.A.,** Induced systemic resistance to disease, in *Biochemical Plant Pathology,* Lamb, C.J., Dixon, R., and Kosuge, T., Eds., Elsevier, Amsterdam, 1984.
33. **Ku, J.H.,** Induced systemic resistance in plants to diseases caused by fungi and bacteria, in *The Dynamics of Host Defence,* Bailey, J.A. and Deverall, B.J., Eds., Academic Press, New York, 1983, 191.
34. **Ku, J.H.,** Plant immunization and its applicability to disease control, in *Innovative Approaches to Plant Disease Control,* Chet, I., Ed., John Wiley & Sons, New York, 1987, 255.
35. **Mercier, J. and Arul, J.,** Induction of systemic disease resistance in carrot roots by pre-inoculation with storage pathogens, *Can. J. Plant Pathol.,* 15, in press.
36. **Overeem, J.C.,** Pre-existing antimicrobial substances in plants and their role in disease resistance, in *Biochemical Aspects of Plant-Parasite Relationships,* Friend, J. and Threlfall, D.R., Eds., Academic Press, New York, 1976, 195.
37. **Stoessl, A.,** Secondary plant metabolites in preinfectional and postinfectional resistance, in *The Dynamics of Host Defence,* Bailey, J.A. and Deverall, B.J., Eds., Academic Press, New York, 1983, 71.
38. **Grainge, M. and Ahmed, S.,** *Handbook of Plants with Pest Control Properties,* John Wiley & Sons, New York, 1988.
39. **Sholberg, P.L. and Shimizu, B.N.,** Use of the natural plant products, hinokitiol, to extend shelf-life of peaches, *Can. Inst. Food Sci. Technol. J.,* 24, 273, 1991.
40. **Dikshit, A., Dubey, N.K., Tripathi, N.N., and Dixit, S.N.,** Cedrus oil—a promising storage fungitoxicant, *Stored Prod. Res.,* 19, 159, 1983.
41. **Maruzzella, J.C., Scrandis, D.A., Scrandis, J.B., and Grabon, G.,** Action of odoriferous organic chemicals and essential oils on wood destroying fungi, *Plant Dis. Rep.,* 44, 789, 1960.
42. **Prasad, K. and Stadelbacker, G.J.,** Effect of acetaldehyde on postharvest decay and market quality of fresh strawberries, *Phytopathology,* 64, 948, 1974.
43. **Stewart, J.K., Aharoni, Y., Hartsell, P.L., and Young, D.K.,** Acetaldehyde fumigation at reduced pressures to control the green peach aphid on wrapped and packed head lettuce, *J. Econ. Entomol.,* 73, 149, 1980.
44. **El Ghaouth, A., Arul, J., and Asselin, A.,** Potential use of chitosan in postharvest preservation of fruits and vegetables, in *Advances in Chitin and Chitosan,* Brine, C.J., Sandford, P.A., and Zikakis, J.P., Eds., Elsevier Applied Science, London, 1992, 440.
45. **Langcake, P.,** Alternative chemical agents for controlling plant disease, *Philos. Trans. R. Soc. London, Ser. B,* 295, 83, 1981.
46. **Ames, B.N., Magan, R., and Gold, L.S.,** Ranking possible carcinogenic hazards, *Science,* 236, 271, 1987.
47. **Lowings, P.H. and Cutts, D.F.,** The preservation of fresh fruits and vegetables, in *Proc. Inst. Sci. Technol. Annu. Symp.,* Nottingham, U.K., July 1981, 52.
48. **Segall, R.H., Down, A., and Davis, P.L.,** Effect of waxing on decay, weight loss and volatile pattern of cucumbers, *Proc. Fla. State Hort. Soc.,* 87, 249, 1974.
49. **El Ghaouth, A., Arul, J., Ponnampalam, R., and Boulet, M.,** Chitosan coating on storability and quality of fresh strawberries, *J. Food Sci.,* 56, 1618, 1991.

50. **El Ghaouth, A., Arul, J., Ponnampalam, R., and Boulet, M.,** Use of chitosan coating to reduce water loss and maintain quality of cucumber and bell pepper fruits, *J. Food Process. Preserv.,* 15, 359, 1991.

51. **El Ghaouth, A., Ponnampalam, R., Castaigne, F., and Arul, J.,** Chitosan coating to extend the storage life of tomatoes, *HortScience,* 27, 1016, 1992.

52. **Allan, C.R. and Hadwiger, L.A.,** The fungicidal effect of chitosan on fungi of varying cell wall composition, *Exp. Mycol.,* 3, 285, 1979.

53. **El Ghaouth, A., Arul, J., Grenier, J., and Asselin, A.,** Antifungal activity of chitosan on two postharvest pathogens of strawberry fruits, *Phytopathology,* 82, 398, 1992.

54. **El Ghaouth, A., Arul, J., Asselin, A., and Benhamou, N.,** Antifungal activity of chitosan on post-harvest pathogens: induction of morphological and cytological alterations in *Rhizopus stolonifer, Mycol. Res.,* 96, 769, 1992.

55. **Hirano, S. and Nagao, N.,** Effect of chitosan, pectic acid, lysozyme and chitinase on the growth of several phytopathogens, *Agric. Biol. Chem.,* 53, 3065, 1989.

55a. **Wilson, C.L.,** personal communication, 1993.

56. **Dilley, D.R. and Wilson, I.D.,** Molecular biological investigations of gene expression attending fruit ripening: current status and future prospects, *HortTechnology,* 2, 294, 1992.

57. **Gray, J., Picton, S., Shabbeer, J., Schuch, W. and Grierson, D.,** Molecular biology of fruit ripening and its manipulation with antisense genes, *Plant Mol. Biol.,* 19, 69, 1992.

58. **Smith, C.J.S., Watson, C.F., Morris, P.C., Bird, C.R., Seymour, G.B., Gray, J.E., Arnold, C., Tucker, G.A., Schuch, W., Harding, S.F., and Grierson, D.,** Inheritance and effects on ripening of antisense polygalacturonase genes in transgenic tomatoes, *Plant Mol. Biol.,* 14, 369, 1990.

59. **Oeller, P.W., Min-Wong, L., Taylor, L.P., Pike, D.A., and Theologis, A.,** Reversible inhibition of tomato fruit senescence by antisense RNA, *Science,* 254, 437, 1991.

60. **King, G.A. and Davies, K.M.,** Identification, cDNA cloning and analysis of mRNAs having altered expression in tips of harvested asparagus spears, *Plant Physiol.,* 100, 1661, 1992.

61. **King, G.A., Hurst, P.L., Irving, D.E., and Liu, R.E.,** Recent advances in the postharvest physiology, storage and handling of green asparagus, *Postharvest News Inf.,* 4(3), in press.

62. **Collinge, D.B. and Slusarenko, A.J.,** Plant gene expression in response to pathogen, *Plant Mol. Biol.,* 9, 389, 1987.

63. **Lamb, C.J., Lawton, M.A., Dron, M., and Dixon, R.A.,** Signals and transduction mechanisms for activation of plant defense against microbial attack, *Cell,* 56, 215, 1989.

64. **Broglie, K., Chet, I., Holliday, M., Cressman, R., Biddles, P., Knowlton, S., Mauvals, C.J., and Broglie, R.,** Transgenic plants with enhanced resistance to the fungal pathogen *Rhizoctonia solani, Science,* 254, 1194, 1991.

65. **Chalutz, E. and Wilson, C.L.,** Postharvest biocontrol of green and blue mold and sour rot of citrus fruit by *Debaryomyces hansenii, Plant Dis.,* 74, 134, 1990.

66. **Janisiewicz, W.J.,** Biocontrol of postharvest diseases of apples with antagonist mixtures, *Phytopathology,* 78, 194, 1988.

67. **Pusey, P.L., Hotchkiss, M.W., Dulmage, H.T., Baumgardner, R.A., Zehr, E.I., Reilley, C.C., and Wilson, C.L.,** Pilot tests for commercial production and application of *Bacillus subtilis* (B-3) for postharvest control of peach brown rot, *Plant Dis.,* 72, 622, 1988.

68. **Wilson, C.L. and Wisniewski, M.E.,** Biological control of postharvest diseases of fruits and vegetables: an emerging technology, *Annu. Rev. Phytopathol.,* 27, 425, 1989.

69. **Wisniewski, M.E. and Wilson, C.L.,** Biological control of postharvest diseases of fruits and vegetables: recent advances, *HortScience,* 27, 94, 1992.

70. **Wilson, C.L.,** Managing the microflora of harvested fruits and vegetables to enhance resistance, *Phytopathology,* 79, 1387, 1989.

71. **Chalutz, E., Droby, S., and Wilson, C.L.,** Mechanism of action of postharvest biocontrol agents, in *Proc. 5th Int. Congr. Plant Pathol.,* Kyoto, Japan, 1988, 442.

72. **Gueldner, R.C., Reilly, C.C., Pusey, P.L., Costello, C.E., and Arrendale, R.F.,** Isolation and identification of iturins as antifungal peptides in biological control of peach brown rot with *Bacillus subtilis, J. Agric. Food Chem.,* 36, 366, 1988.

73. **Wilson, C.L., Franklin, J.D., and Pusey, P.L.,** Biological control of *Rhizopus* rot of peach with *Enterobacter cloacae, Phytopathology,* 77, 303, 1987.

74. **Wisniewski, M.E., Wilson, C.L., Chalutz, E., and Hershberger, W.,** Biological control of postharvest diseases of fruit: inhibition of *Botrytis* rot on apples by an antagonistic yeast, *Proc. Elec. Microsc. Soc. Am.,* 46, 290, 1988.

74a. **El Ghaouth, A. and Wilson, C.L.,** personal communication, 1993.

Chapter 2

The Microbial Ecology of Fruit and Vegetable Surfaces: Its Relationship to Postharvest Biocontrol

Harvey W. Spurr, Jr.

CONTENTS

I. INTRODUCTION

As implied by the title, you may assume there is a relationship between microbial ecology of plant surfaces and biological control of disease on these surfaces. Bacterial, fungal, and other plant pathogens must contact and pass through these surfaces to infect and cause disease — pathogenesis. This chapter will focus primarily on foliar, fruit, and vegetable surfaces in aerial preharvest environments. Fruits and vegetables grow in a wide range of environments, and those in aerial environments are in proximity to foliage — buds, leaves, stems, and flowers — which share this environment and influence biological events including disease initiation and development. This aerial environment is different from but not unrelated to the soil environment which influences the microbial ecology of bulb, stem, and root surfaces. Thus, aerial fruit and vegetables have microflora and diseases which differ from cucumbers on the soil and potatoes under the soil. The reader should reflect on how environmental conditions associated with fruit and vegetable production impact (alter) plant surface microbial ecology and plant disease. Hereafter, the word fruit will be used and understood to include vegetables.

Microbial ecology is the science devoted to the study of microorganisms and their life and death struggles or interactions in microenvironments or in their habitat. One microenvironment is found on a peanut leaf surface and another on a peach fruit surface and so forth. The leaf surface is referred to as the phylloplane (pronounced with a short i: fill-o-plane). This is a two-dimensional surface or plane. Currently the term phyllosphere is used more because it includes the third dimension, the surrounding space. In this space surrounding plant surfaces microbial interactions occur. Here bacteria, fungi, algae, pollen, exudates, pollutants, and other constituents interact under various conditions of humidity, temperature, and light. The development of microbial communities in the phyllosphere is a facet of microbial ecology which impacts foliar pathogens and alters disease development.

Plant pathologists have studied extensively the relationship of moisture, temperature, and other environmental factors to the dispersal and growth of bacterial and fungal pathogens. Thus, plant pathologists are microbial ecologists of sorts. However, plant pathologists seldom concern themselves with the "other microorganisms", the nonpathogens, in the same microenvironment with pathogens. Their focus has been on pathogen-environment interactions in relation to pathogenesis. Ecology and microbial ecology are uncommon words in pathological verbiage and textbooks. This is because the traditional agricultural focus has largely excluded ecological principles. With the realization that pathogens in contact with plant surfaces are not alone but part of a community of microorganisms interacting in a microenvironment or habitat and playing a role (filling a niche), ecology emerges with its potential for understanding biological control. This realization has alerted many plant pathologists to a new awareness of microbial ecology and the potential of using microbial agents — nonpathogenic microbes — to manage or control plant pathogens.

To date, microbial biological control (biocontrol) of plant disease is not a practical, economical method. It is the purpose of this chapter to examine why this is true and how the current situation relates to the research approaches used and microbial ecology.

II. TWO APPROACHES TO MICROBIAL BIOCONTROL

Plant disease researchers look to biological methods for disease control when traditional disease control methods have failed, have undesirable effects, or are not available. Thus when host plant resistance, fungicide application, or cultural practices do not provide effective, economical control, biocontrol may be considered. Whereas biocontrol is attractive as a user safe, environmentally sound method, it has a poor track record. Few examples of practical, successfully implemented systems can be cited or used as guides for developing additional biocontrol systems. Why is this and what can be done to change the situation? Many scientists have stated that development of biocontrols must be based on an enlightened rather than an empirical approach. That is, a system based on knowledge of mechanisms is more likely to succeed.

A. THE EMPIRICAL APPROACH

The empirical approach, often called the "trial and error" method and referred to as "screening", relies on a simple premise: some chemicals are fungicidal and some plants are disease resistant. Therefore, if numerous chemicals or breeding lines are screened, some effective candidates will be identified for additional testing and development.

Using the empirical approach or screening to find control agents does not imply that knowledge of mechanisms is unimportant. Available knowledge often is or may be incorporated into the screening process either in a primary test or in secondary, more advanced, tests. Thus, knowledge is often used to design more sophisticated or efficacious screening and testing systems. Plant pathologists have developed many bioassays as screening tests to identify microbial antagonists to plant pathogens. These tests vary from *in vitro* laboratory petri dish bioassays to *in vivo* greenhouse and field evaluations of disease control.

The use of this empirical approach was called the "silver bullet" approach by Spurr and Knudsen.[1] This analogy was used because the probability of finding one microorganism to control one pathogen was as likely as killing a werewolf with a silver bullet. The empirical approach was highly successful in the search for fungicides and disease-resistant plants. Some of this success can be attributed to screening large numbers of candidates. In the case of chemicals, thousands of chemicals have been tested resulting in a large repository of information — knowledge — on relationships between chemical properties and fungicidal activity. Large numbers of plants — breeding lines — have also been screened. Resistance has been identified by obtaining plants from various sources including isolated and separated habitats where ecological pressures resulted in different genotypes.

By contrast, large numbers of microbes have not been screened for biocontrol in individual research projects. Most researchers have screened fewer than 100 microbial isolates for biocontrol activity. The bioassays most often used for screening microbes are designed to detect antagonism resulting from antibiotic production in agar media in petri dish tests. In these tests selected isolates are grown on plates adjacent to bacterial or fungal pathogens. These bioassays do not assess the potential of microbial isolates to control pathogens by other mechanisms such as competition or hyperparasitism. Thus, plant pathologists have largely utilized antagonism-based laboratory screening procedures to select antibiotic-producing microbial candidates for greenhouse and field tests. Several recent reviews have emphasized the need to develop more realistic and sophisticated techniques for identifying microbial biocontrol agents.[2-4]

Many of the microbial antagonists discovered in petri dish bioassays control disease in greenhouse tests. The usual method, for example, is to spray a candidate microbe on foliage and then spray-inoculate a pathogen. The treated plants are then incubated under conditions of temperature and humidity which favor disease development. These conditions often favor the growth of the selected antagonist on the foliar surface also and result in disease control. This test situation involves host plants but these plants are really functioning as does an agar medium in a petri dish bioassay. Nevertheless, positive disease control in these tests encourages researchers to attempt a similar test under field conditions.

Field tests require more resources, including time and effort, and almost always result in disease control in the range from 0 to 70%. The 70% control level is difficult to surpass. Also, biocontrol field tests are not consistent in efficacy from year to year. At this point most researchers become discouraged and "drop out". In a survey of abstracts for the 1990 Annual Meeting of The American Phytopathological Society, Spurr et al.[5] determined that in 62% of the abstracts reporting on biocontrol, the researcher used the "silver bullet" or empirical approach. Clearly, the progress to date or lack of progress, should emphasize to plant pathologists researching microbial control of plant disease that an enlightened approach or knowledge-based approach is needed.

B. THE ENLIGHTENED APPROACH

As stated, the progress in developing plant disease control including chemical control, host resistance, and cultural practices was based on trial and error, the application of empirical methods. In other words, this is the "try this and ask why later" approach. This has resulted in many successful practices and has eliminated most of the large-scale disease epidemics in agriculture. The knowledge of disease control mechanisms — such as why this cultivar is resistant to this disease — remains unknown in most instances. Success has not been accompanied by a similar gain in knowledge. However, it has been stated and restated that disease control efficacy with chemicals, host resistance, and cultural practice would benefit or improve when a larger knowledge base became available. This has not happened.

In the current situation with microbial biocontrol, which has a largely unsuccessful record, the premise is that knowledge must precede control.

Thus, the "enlightened approach" to microbial biocontrol depends on developing a large or sufficient knowledge base — knowledge of microbial ecology — which will be used to "reason out" successful microbial disease control. Certainly, as everyone knows from problem solving in school exercises, knowledge is the "key to success". Looking at society's problems we must admit that it takes a lot of knowledge to solve complex problems, and the opportunities and resources available for expanding our knowledge are not sufficient.

Nonetheless, plant pathologists have over the years emphatically stated they could improve disease control if more knowledge were available — more basic knowledge on disease physiology, on epidemiology, on host genetics, on mechanisms of fungicidal toxicity, and now on microbial ecology. It is difficult to argue with such a statement. Thus, it was concluded by the Committee on Biological Control Research Needs and Priorities in Plant-Microbe Interactions in Agriculture[6] that: "To achieve microbial biocontrol — manipulation of plant associated microbial communities to diminish the harmful activities of plant pathogens — we need to understand the structure, development dynamics, and regulation of these communities." In other words, they conclude that prior to succeeding with microbial biocontrol, the microbial ecology knowledge base must be considerably enlarged.

With what does this conclusion leave us? Does it mean that numerous plant pathologists pursuing microbial biocontrol via the empirical approach are guaranteed to fail? Does it mean that all of these scientists need to restructure their research to pursue some facet of microbial ecology? Does this mean that industries employing the empirical approach to microbial biocontrol should stop because they have little chance to succeed? It would appear that if everyone engaged in microbial biocontrol research were to switch to the enlightened approach, there would be a radical change in research activity. Perhaps the question should be stated another way. Will a large expansion in the microbial ecology knowledge base result in the development of practical microbial biocontrol?

Undoubtedly, all plant pathologists vigorously pursuing the empirical approach to microbial biocontrol are not about to switch into microbial ecology and the enlightened approach. What is happening with some regularity, is that enthusiasm generated from successful laboratory experiments is being deflated by field experimentation. The "up one minute and down the next" should cause a reexamination on the part of these scientists. The "What went wrong?" and "How can we improve performance?" kind of questions are generated. Also, did our microbial agent survive and grow in its new habitat following its application in the field? These are microbial ecology questions which gradually change the empirical approach, in small steps, toward a "modified empirical approach" which is closer to the enlightened approach. It is not likely that a large movement to basic microbial ecology research will occur. One reason is that it is important for plant pathologists to present and preserve the impression that they are working on projects that lead directly or indirectly to disease control in the not too distant future. This strong, innate drive to be associated with developing disease controls — an economic rationale for our research activity — seems more acceptable, more salable than doing research unrelated to immediate or direct, short-term societal benefit. Our best chance for developing practical microbial biocontrol depends on successfully integrating microbial ecology research with the empirical research now in progress.

III. PLANT SURFACES: A MAGNIFIED VIEW

Plant disease control has been compared with military operations because scientists envision the enemy as pathogenic spores, the battleground as a plant surface, and the defensive force as those elements which can be assembled and managed to fight pathogenic spores. This analogy serves the purpose of creating a picture of the unseen surface constituents and the strategies and tactics for their management. Here the focus is on the battlefield, the plant surface or specifically the phylloplane. What is it really? How does it look?

Experience with plants and food have shown us many of the differences. We are aware of the superficial differences that exist between lettuce leaves and peach fruit. The specific differences are what we need to know. These differences are defined under the broad categories of physical, chemical, biological, and environmental characteristics. This may appear elementary but a consideration of these elements and their impact is often overlooked; that is, many scientists pursuing biological control do not make a critical examination of the battlefield. The following are a few general observations about phylloplane surfaces and the phyllosphere. Detailed reviews are available on plant surfaces[7] and their microbiology.[8]

A. PHYSICAL CHARACTERISTICS

The physical surface depends on the size and shape of epidermal cells which vary considerably. What appears as a smooth surface to our touch and vision may be seen, when magnified, as rough terrain for microorganisms. We are familiar with epidermal and guard cells but are less acquainted with special function cells such as trichomes, or leaf hair cells. There may be glandular and/or nonglandular trichomes present, branched and unbranched in structure. Also, the density of trichomes varies from dense forests to sparse deserts. Populations of the nitrogen-fixing bacterium *Beijerinckea* sp. on cotton leaves of *B. varalaxmi* showed that the majority of the microbes were on the trichomes.[9] This leads to the question: Would the same biocontrol tactic be equally effective for controlling brown rot on smooth peaches as on pubescent peaches?

B. CHEMICAL CHARACTERISTICS

Glandular trichomes produce a range of chemicals which influence the growth and development of microbes. Numerous chemicals are produced in glandular hairs of different plant species. These chemicals are exuded onto the surface where their activities include repelling insects and inhibiting spore germination.[7,10,11] These activities are in addition to those from other surface chemicals exuded from or leaking from epidermal cells. Other chemicals come from pollen and pollutants in the atmosphere. A large number of chemicals found on leaf surfaces leak from epidermal cells in association with plant growth and development. In addition, various pesticides may be applied at different times. These dramatically change the chemical composition of the surface.

C. BIOLOGICAL CHARACTERISTICS

What are the microbes of the phyllosphere communities? They are the bacteria and fungi growing on plant surfaces. Studies have enumerated many species. It seems evident that the plant-associated microbes occur in a succession over time. For example, Blakeman[12] described a general succession in microbial population development on annual leaves from emergence to senescence. The microbial population of leaves is dynamic in composition of species, numbers, and habitat. Some microbes are surface inhabitants, "epiphytes"; and some parasitize tissues, "endophytes". Bacteria and yeasts often are more prevalent early in the growing season, and saprophytic filamentous fungi, later in temperate regions. Most assessments of microbial populations on surfaces are incomplete or fragmentary in that not all microbes present over time are identified. Also, what constitutes a microbial community on plant surfaces has not been defined. Many studies describe microbial species and their numbers over time during the growing season, but little specific information is available to describe what constitutes a microbial community on a plant surface. What are the makeup and roles of community members and how are they influenced by physical, chemical, or other events? In "more studied habitats", such as rocks in streams, the development of biofilms composed of different microbial species has been described and characterized.[13]

D. ENVIRONMENTAL CHARACTERISTICS

The driving forces of temperature, moisture, and radiation are often overlooked or taken for granted. A severe rainstorm, a drought, or a cool growing season all impact plant growth and development, and likewise, all influence plant-associated microbial communities and plant disease development. For instance, rain will wash some chemicals and microbes from leaf surfaces. At the same time it provides moisture for microbial growth and development including both pathogens and antagonistic nonpathogens. Thus, environmental factors drive the dynamic nature of microbial communities on surfaces.

This brief description of physical, chemical, biological, and environmental factors impacting the phyllosphere was intended to enlarge and magnify the reader's view of leaf surfaces. Also, it was intended to provide an appreciation for the complexity of the microbial ecosystem associated with these habitats. This ecosystem diversity and complexity continues to be underestimated by many plant pathologists. Knowing and better understanding this system appears to many observers as crucial to achieving successful biological control of plant disease.

IV. THE MICROBIAL STATE AT HARVEST

The microbial population of a fruit will have developed to a "state" at harvest. This state is the sum of physical, chemical, biological, environmental, and cultural inputs during the growing season. It is the state of the microbial population at harvest which is a major factor in determining postharvest rot. As one focuses on postharvest rot problems, the state at harvest is usually overlooked or considered in a cursory manner. Do we ask ourselves basic questions about the state such as: Is the state of healthy appearing fruit the same for all fruit harvested in one season? Is the state of all fruit varieties the same or similar at harvest? Essentially the question is: How much variation exists in the state at harvest and how does the variation impact rot development and control?

We cannot specifically answer these questions. However, we know that fruit produced for the commercial market is subjected to numerous cultural practices which include various pesticide applications and is subjected to variations in weather conditions. All of these factors alter microbial populations and contribute to differences in the state at harvest.

The microbial state at harvest then consists of numerous microorganisms forming a population. This population is comprised of mixtures of species which we refer to as a community. The species of the community include bacteria, fungi, yeasts, and other microorganisms. Estimates indicate that microorganisms fluctuate in number between 10^3 and 10^7 per square centimeter of surface area and that this may be an evenly or unevenly distributed population. Thus, dense mixtures of microorganisms may occur in some areas and sparse or no populations, in others. Some of these microorganisms, approximately 5% are pathogens. A 5% pathogen population may be more than sufficient to cause severe rot or unacceptable levels of rot under rot-favorable conditions. These pathogens may be living on tissue saprophytically, parasitically, or pathogenically. If pathogens have infected the tissue, development stages may vary from nonvisible, latent (quiescent) stages to highly visible advanced stages. Some of the microbial population is on the surface, perhaps associated with hairs or epidermal cells, and some is below the surface in substomatal cavities or within epidermal and parenchymal cells.

At harvest, the state of the microbial communities associated with fruit is complex in composition, numerically variable, and generally unknown. The state is not evaluated except for visible disease caused by major pathogens. The microscopic nature of the microbial state of fruit at harvest precludes a simple assessment of its character. We know it is composed of numerous bacteria, yeast, and filamentous fungi; and this population of microorganisms is the sum of the natural and artificial inputs or events which occurred during the growing season. The point is that we know more about 5% of the population — the pathogens — than the other 95% — the nonpathogens — which are a primary hope for disease control. Again, the development of more knowledge of the microbial state at harvest and learning to manage this state are prerequisite to microbial biocontrol.

V. HARVESTING AND HANDLING: IMPACT ON THE MICROBIAL STATE

As indicated, the microbial state at harvest is a major factor in determining postharvest rot. A second major factor is harvesting and handling. The transition from field to shipping or storage subjects fruit to bruising, wounding, washing, dipping, heating, and cooling in various combinations which are determined by industry. Bruising and wounding during harvesting and handling have long been recognized as predisposing fruit to rot by creating infection sites that are readily colonized by rot pathogens. Also, handling may serve to spread or redistribute pathogen inoculum (spores). It is important to realize that all microflora present, pathogenic and nonpathogenic, contaminate processing equipment and are redistributed over the fruit surface. To date, the focus during harvesting and handling of fruit has been on minimizing the impact of rot pathogens. The opportunity to maximize the impact of nonpathogens that contribute to rot control should become a focal point.

VI. WHAT IS KNOWN ABOUT THE MICROBIAL STATE AT HARVEST

The previous paragraphs described in general terms the importance of the microbial state of fruit at harvest to postharvest development of rot. There are few publications which describe microbial populations of fruit. Wilson et al.[14] made this point when he stated, "Although considerable research on the microbiology and microecology of the phylloplane exists, little information exists for fruit and vegetable surfaces." Therefore, an examination of what is known of the microbial state of leaves for comparison to that of fruit may be helpful.

When leaves or flowers emerge from buds, the tissues undergo colonization by microorganisms. These microorganisms move from seed, soil, air, and plant stems and buds, by wind and splashing to the new tissues.[15,16] Early arrivals may colonize suitable sites and survive as residents through the growing season until senescence and sometimes through the decay process.

Blakeman[12] described bacteria as early colonizers of leaves when nutrient levels on surfaces are low. As nutrients increase from cell leakage, pollen, and insect honeydew, the numbers of yeast and filamentous fungi increase. Total numbers of microorganisms tend to increase with time after emergence of tissue until a state is reached where the population growth stabilizes. This state may relate to the "carrying capacity" which is dependent on nutrients, space, moisture, and competition. This state is dynamic since fluctuations in the populations occur in response to driving factors such as rain and pesticide application. Microbial life cycles in the phyllosphere vary with species. For example, *Alternaria* once established appears to survive for the season, whereas *Cladosporium* spores can germinate, parasitize, and sporulate in a 3-d period in leaves.[17-19] These fungi grow endophytically as parasites whereas most of the bacteria, yeasts, and some filamentous fungi grow epiphytically — on the surface.

A considerable diversity in microbial species colonize foliage over the season. In a comprehensive study of epiphytic populations on McIntosh apple leaves, Andrews and Kenerley[20,21] and Andrews[22] observed numerous bacterial species, yeasts, and filamentous fungi. These varied in number from season to season. It was apparent that differences in the frequency and diversity of microflora varied with leaf position in the apple tree canopy and were altered by pesticide application. Thus, apple pesticide programs have a substantial impact on non-target epiphytic microorganisms. Unfortunately, apple fruit surfaces were not sampled in this study.

Andrews and Kenerley[20] reported that the filamentous fungi *Alternaria*, *Cladosporium*, and *Coniothyrium*; and the yeasts *Aureobasidium* and *Sporobolomyces* were the dominant microflora. The microbial species found in the population of microflora on apple leaves in Wisconsin was similar to microflora on tobacco leaves in North Carolina;[18] tobacco leaves in Malawi, Africa;[23] and leaves of many plant species growing in temperate zones or regions. Thus, numerous studies have shown considerable diversity in species of yeasts and fungi colonizing leaves, but also considerable similarity over the temperate regions. Generally 20 to 40 species of yeast and fungi may be identified in a report. Thus, *Alternaria* and *Cladosporium* are always among the dominant fungi. *Aureobasidium* and *Sporobolomyces* are among the dominant yeasts. However, fungi such as *Epicoccum* and yeasts such as *Rhodotorula* are nearly always reported as present but to a lesser extent. Does this mean they are less important or do they have different impacts?

What about bacterial species colonizing leaves? In the study of apple leaves Andrews and Kenerley[20] found significant numbers of fluorescent pseudomonads. This is typical of leaves based on many surveys and reports.[24] Most bacteria are characterized on different growth media when aqueous extracts are dilution plated. Some data were obtained by using a leaf imprint technique. This is an approach used by many researchers. The results indicate that species of bacteria have not been identified to the extent that fungi have been identified. The problem is that the skills, techniques, and time required to adequately evaluate bacterial populations are more demanding and consume more resources. Thus, we know that bacteria are present and are diverse in species composition, but we have less information on bacteria on leaf surfaces. New, more rapid techniques, such as identifying bacteria based on fatty acid methyl ester (FAME) profiles are changing this.

Another aspect of this problem of defining microbial leaf surface populations resides with the skills, interest, and objectives of individual researchers. For example, Murty[7] studied extensively the bacterial flora of cotton leaves in India. The bacterium *Beijerinckea* sp. was identified as an extensive colonizer of cotton leaves. This bacterial species fixes nitrogen from the air and was estimated to fix from 1 to 10% of the nitrogen used by cotton plants during a growing season.

Although these nitrogen-fixing bacteria grow on cotton leaf hairs, they are just as frequent on the leaf surfaces of hairless varieties of cotton fixing an equivalent amount of nitrogen. The association of these bacteria with leaf hairs may be described as intimate because when radioactive phosphorus was applied to cotton roots, it was detected in the leaf hair bacteria. This also confirms the importance of soil-root activity to foliar activity.

In reading this account of the relationship of *Beijerinckea* sp. to cotton leaves it may seem that this is the only microorganism present. However, as Murty states this is one species identified from the general bacterial population and although it may be considered a major species in the cotton leaf phyllosphere, the activities of other bacteria are unknown. Also, the activities of yeast and fungi in the systems were not described.

Most often *Pseudomonas* and *Erwinia* species have been described as bacterial epiphytes of foliage. These are Gram negative; however, it is not unusual for Gram positive, spore-forming bacteria to also be isolated and identified. In an extensive study of bacterial flora of peanut leaves, Elliott and Spurr[25] have identified and quantified 18 bacterial species. This is one of the most detailed accounts of the diverse number of bacterial species on leaves of one plant species. Many of the species such as *Agrobacterium* have not been considered previously to be leaf colonizers. Elliott[26] describes factors which could explain the distribution of bacterial populations on leaf surfaces. Both normal and log-normal population distributions of epiphytic bacteria occur depending on the species in the population, time of sampling, available food, and many other factors. This research emphasizes the need for more directed resources to gain a full understanding of microbial populations in the phyllosphere.

VII. THE MICROBIAL STATE AND POSTHARVEST BIOCONTROL

What is the microbial state on fruit surfaces such as apple, peach, or grapefruit at harvest? It is clear that fruit ready for harvest has a microflora composed of bacteria, yeasts, and fungi as does the surrounding foliage. However, it is unfortunate that the descriptions of fruit microfloral populations have been less rigorously studied than foliage. Although the microbial population of apple leaves at harvest has been described extensively, the composition of the adjacent apple fruit surface has not been fully described.

Over the past decade, several investigations were initiated to explore the potential of microbial antagonism as a means to control several fungi which cause postharvest fruit rot. This interest was stimulated by a need to control rot and the desire to reduce the risks to consumers posed by postharvest applications of fungicides.

One of the first of recent reports was by Pusey and Wilson,[27] who demonstrated the control of brown rot of peach by an isolate of *Bacillus subtilis*. This bacterium produced an antibiotic substance which was toxic to the pathogen *Monilinia fructicola*. It was not effective against *Rhizopus* rot which increased when *M. fructicola* was controlled. Soon after, research with apple rot control resulted in the isolation of another antibiotic-producing bacterium, *Pseudomonas cepacia*[28] which controlled blue mold and gray mold of apple and pear. This was followed quickly by the isolation of several nonantibiotic-producing yeasts which effectively controlled rots. Chalutz et al.[29] reported the broad activity of *Pichia guilliermondii* (US-7) (formerly *Debaryomyces hansenii*) for control of green mold, blue mold, and sour rot of citrus. Roberts[30] reported effective control of gray mold of apple by *Cryptococcus laurentii*. Thus, over the decade, investigations moved from using antibiotic-producing strains of bacteria to antagonistic yeasts, probably site competitors, for control of numerous fruit rots. This research is largely summarized in the proceedings of a workshop sponsored by the Agricultural Research Service, U.S. Department of Agriculture and U.S./Israeli Binational Agricultural Research Fund.[31]

This brief overview of recent postharvest rot research emphasizes the common approach used in these investigations. Each investigator begins with the concept of finding suitable microbial antagonists to control postharvest rot. The search for these antagonists begins with fruit surfaces where there is no rot. Thus, by *a priori* logic based on the knowledge that fruit surfaces host microorganisms, those microorganisms residing where there was no rot were likely to antagonize rot-causing microorganisms. These surfaces were subjected to aqueous extraction, followed by plating and isolation of pure strains of bacteria and yeast. Filamentous fungi were ignored. The isolates were then screened *in vivo* in wound sites for competitive inhibition of rot pathogens. In summary, although as many as 200 isolates from one type of fruit surface were isolated and screened in an effort to find antagonists, these investigations did not enhance our knowledge of microbial ecology of fruit surfaces. There were no detailed reports of the microbial composition of fruit surfaces.

These investigations represent the empirical or "silver bullet" approach to biological control. This does not mean there is no value in the data and the knowledge generated. Some new knowledge was generated. If this research were successful, a large number of postharvest fruit rots would now be commercially controlled by antagonists.

VIII. ENHANCEMENT AS A MEANS TO POSTHARVEST BIOCONTROL

What do you do when you realize that your microbial antagonist has failed to provide sufficient control of postharvest rot? Most researchers look around for some means to enhance the antagonistic activity. Enhancers are chemicals that serve as a food base for antagonistic microbes or are selectively toxic or fungicidal to rot pathogens. Thus, one of the early observations made by Chalutz et al.[29] was that their US-7 isolate of *P. guilliermondii* was more resistant to the fungicides thiabendazole and imazalil than *Penicillium digitatum*, the pathogen causing green mold of citrus fruit. Therefore, the fungicide-tolerant antagonist could be combined with a fungicide — probably at a lowered concentration, to enhance the spectrum of control and efficacy to reach the goal of commercially acceptable rot control.

A more unusual enhancement of antagonism by yeasts resulted from the addition of calcium salts such as 2% (w/v) calcium chloride to cell suspensions of antagonistic yeast isolates.[32] Apparently the calcium chloride causes a reduction in germination and germ tube elongation of certain postharvest pathogens such as *Penicillium* spp. and *Botrytis cinerea*.

The addition of a food source to selectively feed microorganisms *in situ* and thereby enhance antagonisms has been long been considered. Morris et al.[33] investigated simple sugars and amino acids as food sources of the bacterial epiphyte and brown spot pathogen of bean leaves, *Pseudomonas syringae*. They sprayed various combinations of sugars and amino acids on bean leaves in the field to determine the impact on epiphytic bacterial populations and the development of bacterial brown spot. They concluded that the application of simple organic compounds to bean leaf surfaces can alter the composition of the bacterial community. Spraying a food source such as glycine on bean leaves reduced disease caused by *P. syringae*. The mechanism was complex and was not immediately apparent such as being directly related to impacting microbial growth.

In recent field tests by Davis et al.,[34] spores of the fungal antagonist *Chaetomium globosum* were spray-applied to apple leaves and fruit which had previously been sprayed with cellulose, a food source for this fungus. The addition of cellulose enhanced the growth of *C. globosum* and reduced the incidence of sooty blotch and flyspeck on apple fruit. In other research originating in this laboratory,[35] foliar amendments of chitin were used to enhance the biological control of peanut leaf spot by a chitinolytic *Bacillus subtilis* isolated from peanut leaves. These experiments demonstrate the potential of selective food substrates to increase the efficacy of fungal antagonists.

IX. CONCLUSIONS

The intensity of the effort to achieve pre- and postharvest control of plant disease by the management of microbial antagonists has steadily increased. Over the past decade the number of publications reporting microbial-pathogen antagonism increased significantly. In 1990 more than 10% of the presentations at the Annual Meeting of The American Phytopathological Society were on biological control research.[5] In spite of the intensity of this focus, practical, reliable biocontrol systems for soilborne, foliar, and postharvest disease do not exist.

In trying to develop an explanation for the elusive nature of "high performance biocontrol" two research approaches were discussed: (1) the usual empirical or "silver bullet" approach followed by most researchers and (2) the never used enlightened approach which assumes that a mastery of microbial ecology must precede the development of effective control systems. Both approaches are dependent on identifying and using the extensive microflora living in association with plants, mostly on plant surfaces. With both approaches the intention is to successfully manage this plant-associated microflora. The empirical approach borders on simplicity because it assumes that one can introduce a single microorganism, whether isolated from the plant surface to be protected or anywhere, back into the system and totally succeed. There are too many pathogens to control, too much surface area to cover, too many environmental factors to confront, and just too much complexity.

On the other hand, the idea that plant pathologists can devote sufficient resources, especially time in years, to making detailed studies of microbial ecology seems unlikely. There is a lack of patience and resources coupled with a strong desire, almost a requirement, to achieve a quick success. Time is not on the side of researchers seeking microbial biocontrol. No one knows how much microbial ecology research is required prior to reaching the goal of being able to effectively manage plant-associated microorganisms. This goal will only be reached if the assumption or premise is correct, i.e., that microflora on plant surfaces can be managed to provide practical, effective disease control.[8]

Perhaps an approach which is between the extremes of empiricism and enlightenment offers more hope? It would not be too difficult to better define the microbial state of fruit surfaces at harvest. For instance, most of the microbes isolated from fruit surfaces are "r" strategists (grow fast, sporulate prolifically, produce and tolerate antibiotics, use many nutrients as food, and tolerate a wide range in environment). Most (nearly all) of the microbial antagonists isolated and tested thus far are r strategists. Microbes that lack the above characteristics are called "K" strategists (avoid intense competition by some advantage, such as grow on low nutrient level or utilize an unusual source of nutrient). These microbes are present most of the time and are less dynamic in population fluctuation.[8,36] Developing a better understanding of K strategists appears to be needed.

The use of enhancement techniques in conjunction with microbial antagonism may be another approach between the extremes. Certainly providing a food source which favors growth, reproduction, and survival of microbial antagonists is an important step in the management process. The use of preharvest, harvest, and postharvest tactics which include physical, chemical, biological, and cultural practice inputs is certainly going to enhance postharvest rot control.[5] Thus, extending the techniques available for enhancement could increase the reliability and practical effectiveness of microbial biocontrol.

Recently Pusey[37] strongly emphasized the need to select microbial antagonists which are commercially acceptable, that is, select microbes which have been used in fermentative processes or have been associated with food products and have a history of being safe. Antibiotic-producing microbes will require greater testing to assure their safety for consumers. Also, many microbes have names that associate them with toxicity to and disease in animals and plants. One of the misfortunes of microbial taxonomy is that everything with the same name is assumed to be similar and pose similar risks. A case can be made for not using species names of promising biocontrol microbes to avoid losing their potential benefits to a psychologically skeptical public. It is unfortunate that such concerns cannot be sidetracked or ignored while scientists concentrate on the basic problems of developing effective microbial control systems. Microbial management has already contributed much to human survival and the quality of life. In time many more important contributions are likely, including the postharvest control of fruit rot.

REFERENCES

1. **Spurr, H. W., Jr. and Knudsen, G. R.,** Biological control of leaf diseases with bacteria, in *Biological Control on the Phylloplane,* Windels, C. E. and Lindow, S., Eds., APS Press, St. Paul, MN, 1985, 45.
2. **Spurr, H. W., Jr.,** Bioassays — critical to biocontrol of plant disease, *J. Agric. Entomol,.* 2, 117, 1985.
3. **Knudsen, G. R. and Spurr, H. W., Jr.,** Management of bacterial populations for foliar disease biocontrol, in *Biocontrol of Plant Diseases*, Vol. 1, Mukerji, K. G. and Garg, K. L., Eds., CRC Press, Boca Raton, FL, 1988, 83.

4. **Andrews, J. H.,** Strategies for selecting antagonistic microorganisms from the phylloplane, in *Biological Control on the Phylloplane,* Windels, C.E. and Lindow, S., Eds., APS Press, St. Paul, MN, 1985, 31.

5. **Spurr, H. W., Jr., Elliott, V. J., and Thal, W. M.,** Managing epiphytic microflora for biocontrol, in *Biological Control of Postharvest Diseases of Fruits and Vegetables,* Wilson, C. L. and Chalutz, E., Eds., Agricultural Research Service, U.S. Department of Agriculture, ARS-92, 1991, 3.

6. Committee on Biological Control Research Needs and Priorities in Plant-Microbe Interactions, Board on Biology, Commission on Life Sciences, National Research Council, *The Ecology of Plant-Associated Microorganisms,* National Academy Press, Washington, D.C., 1989, 34.

7. **Juniper, B. E. and Jeffree, C. E.,** *Plant Surfaces,* E. Arnold, London, 1983, 93.

8. **Campbell, R.,** *Plant Microbiology,* E. Arnold, London, 1985, 13.

9. **Murty, M. G.,** Phyllosphere of cotton as a habitat for diazotrophic microorganisms, *Environ Microbiol.,* 48, 713, 1984.

10. **Cutler, H. G., Severson, R. F., Cole, P. D., Jackson, D. M., and Johnson, A. W.,** Secondary metabolites from higher plants, their possible role as biological control agents, in *Natural Resistance of Plants to Pests, Roles of Allelochemicals,* Green, M. B. and Hedin, P. A., Eds., American Chemical Society Symposium No. 296, Washington, D.C., 1986, 178.

11. **Spurr, H. W., Jr.,** New directions in biological control: alternatives for suppressing agricultural pests and diseases, in *UCLA Sym. Mol. Cell. Biol., New Series,* Vol. 112, Baker, R. and Dunn, P., Eds., Alan R. Liss, New York, 1990, 271.

12. **Blakeman, J. P.,** Ecological succession of leaf surface microorganisms in relation to biological control, in *Biological Control on the Phylloplane,* Windels, C. E. and Lindow, S., Eds., APS Press, St. Paul, MN, 1985, 6.

13. **Costerton, J. W.,** Direct ultrastructural examination of adherent bacterial populations in nature and pathogenic ecosystems, in *Current Perspectives in Microbial Ecology,* Klug, M. J. and Reddy, C. A., Eds., American Society Microbiology, Washington, D.C., 1984, 115.

14. **Wilson, C. L. and Wisniewski, M. E.,** Biological control of postharvest diseases of fruits and vegetables: an emerging technology, *Annu. Rev. Phytopathol.,* 27, 425, 1989.

15. **Leben, C.,** Epiphytic microorganisms in relation to plant disease, *Annu. Rev. Phytopathol.,* 3, 209, 1965.

16. **Leben, C.,** The bud in relation to the epiphytic microflora, in *Ecology of Leaf Surface Micro-Organisms,* Preece, T. F. and Dickinson, C. H., Eds., Academic Press, New York, 1971, 117.

17. **Dickinson, C. H.,** Biology of *Alternaria alternata, Cladosporium cladosporoides* and *C. herbarum* in respect of their activity on green plants, in *Microbial Ecology of the Phylloplane,* Blakeman, J. P., Ed., Academic Press, 1981, 169.

18. **Spurr, H. W., Jr. and Welty, R. E.,** Characterization of endophytic fungi in healthy appearing leaves of *Nicotiana* spp., *Phytopathology,* 65, 417, 1975.

19. **Spurr, H. W., Jr.,** Parasitic cycle of *Cladosporium cladosporoides* in healthy tobacco leaf tissue, *Proc. Am. Phytopathol. Soc.,* 4, 142, 1977.

20. **Andrews, J. H. and Kenerley, C. M.,** The effects of a pesticide program on non-target epiphytic microbial populations of apple leaves, *Can. J. Microbiol.,* 24, 1058, 1978.

21. **Andrews, J. H., Kenerley, C. M. and Nordheim, E. V.,** Positional variation in phylloplane microbial populations within an apple tree canopy, *Microb. Ecol.,* 6, 71, 1980.

22. **Andrews, J. H.,** Effect of pesticides on non-target micro-organisms on leaves, in *Microbial Ecology of the Phylloplane,* Blackman, J. P., Ed., Academic Press, London, 1981, 283.

23. **Norse, D.,** Fungal populations of tobacco leaves and their effect on the growth of *Alternaria longipes, Br. Mycol. Soc.,* 59, 261, 1972.

24. **Hirano, S. S. and Upper, C. D.,** Bacterial community dynamics, in *Microbial Ecology of Leaves,* Andrews, J. H. and Hirano, S. S., Eds., Springer-Verlag, New York, 1991, 271.

25. **Elliott, V. J. and Spurr, H. W., Jr.,** Characterization of bacterial flora of peanut foliage by computer-aided fatty acid methyl ester analysis, *Abstr. 5th Int. Symp. Microbiol. Phyllosphere,* 35, 35, 1990.

26. **Elliott, V. J.,** A statistical model explaining the development of log-normal population distributions of epiphytic bacteria, *Phytopathology,* 84, 1994, in press.

27. **Pusey, P. L. and Wilson, C. L.,** Postharvest biological control of stone fruit brown rot by *Bacillus subtilis, Plant. Dis.,* 68, 753, 1984.

28. **Janisiewicz, W. J. and J. Roitmann,** Biological control of blue-mold and grey-mold on apples and pears with *Pseudomonas cepacia, Phytopathology,* 78, 1697, 1988.

29. **Chalutz, E. and Wilson, C. L.,** Biocontrol of green and blue mold and sour rot of citrus by *Debaryomyces hansenii, Plant Dis.,* 74, 134, 1990.

30. **Roberts, R. G.,** Postharvest biological control of gray mold of apple by *Cryptococcus laurentii, Phytopathology,* 80, 526, 1990.

31. **Wilson, C. L. and Chalutz, E.,** *Biological Control Postharvest Dis. Fruits Vegetables, Workshop Proc.,* Agricultural Research Service, U.S. Department of Agriculture, ARS-92, 1991, 324.

32. **McLaughlin, R. J.,** A review and current status of research on enhancement of biological control of postharvest diseases of fruit by use of calcium salts with yeasts, in *Biological Control Postharvest Dis. Fruits Vegetables, Workshop Proc.,* Wilson, C. L. and Chalutz, E., Eds., Agricultural Research Service, U.S. Department of Agriculture, ARS-92, 1991, 184.

33. **Morris, C. E. and Rouse, D. J.,** Role of nutrients in regulating epiphytic bacterial populations, in *Biological Control on the Phylloplane,* Windels, C. E. and Lindow, S. E., Eds., St. Paul, American Phytopathological Society, 1985, 63.

34. **Davis, R. F., Backman, P. A., Rodriguez-Kabana, R. and Kokalis-Burelle, N.,** Biological control of apple fruit diseases by *Chaetomium globosum* formulations containing cellulose, *Biol. Control,* 2, 118, 1992.

35. **Kokalis-Burelle, N., Backman, P. A., Rodriguez-Kabana, R., and Ploper, L. D.,** Potential for biological control of early leafspot of peanut using *Bacillus cereus* and chitin as foliar amendments, *Biol. Control,* 2, 321, 1992.

36. **Andrews, J. H.,** Relevance of r- and K-theory to the ecology of plant pathogens, in *Current Perspectives in Microbial Ecology,* Klug, M. J. and Reedy, C. A., Eds., American Society of Microbiology, Washington, D.C., 1984, 710.

37. **Pusey, P. L.,** Antibiosis as mode of action in postharvest biological control, in *Biological Control Postharvest Dis. Fruits Vegetables, Workshop Proc.,* Wilson, C. L. and Chalutz, E., Eds., Agricultural Research Service, U.S. Department of Agriculture, ARS-92, 1991, 127.

Chapter 3

Strategies for the Isolation and Testing of Biocontrol Agents

Joseph L. Smilanick

CONTENTS

I. INTRODUCTION

Regulatory and public concerns about health risks associated with the ingestion of fungicide residues have reduced the number of these compounds available.[1] These concerns arose although the existence of long-term risk from synthetic pesticides is uncertain[2,3] and poorly supported when quantified.[4,5] Another threat to their continued use is the progressive reduced efficacy of many postharvest fungicides due to chemical resistance among these pathogens.[6] Furthermore, the markets for postharvest fungicides are relatively small, and it has become difficult to justify the costs of new registrations or even maintaining registrations for synthetic chemicals. These issues have stimulated interest in the development of alternative strategies to fungicides to control postharvest diseases of fresh fruits and vegetables. Biocontrol of postharvest diseases by microbial antagonists is among alternative decay control techniques.[1]

Janisiewicz[7,8] and Pusey[9] discussed strategies to develop postharvest biocontrol, and this subject has been addressed in general reviews of this area.[10-13] Strategies to develop phyllosphere biocontrol often are relevant because approaches employed in this field are similar to those used to control postharvest diseases, namely, preemptive colonization of the host by the biocontrol agents before pathogens arrive.[14-18]

Wisniewski and Wilson[11,13] described the characteristics of an "ideal antagonist" for postharvest disease control: genetically stable; effective at low concentrations in many pathogen-host combinations; able to survive adverse environmental conditions such as low temperatures and controlled atmosphere storage; not fastidious in its nutrient requirements; amenable to inexpensive production and formulation with a long shelf life; easy to dispense; resistant to pesticides; compatible with commercial handling practices; does not produce metabolites that are deleterious to human health; and nonpathogenic to the host commodity. Additional concerns are that antagonists must not colonize or pathogenize nontarget plants or animals, especially humans. The purpose of this review is to discuss strategies to find antagonists that have these attributes.

The postharvest fungicides employed today, about 20 in number,[19,20] are the result of screening many candidate compounds for efficacy. In the development of these fungicides, effective compounds remained candidates for commercialization if they were patentable and stable, manufacture was feasible, and a demand for their activity existed. Environmental and animal toxicological properties of these compounds were determined for regulatory purposes, the quantity of residues reaching consumers was estimated, and the resulting information was used in a risk-benefit analysis before a registration was granted. The biochemical mode of action of approved compounds (e.g., benzimidazoles), although useful and important information, was often not known until after commercialization.[21] Development culminated when prescribed amounts of one or more of these relatively purified, approved compounds was applied commercially. Chemical control strategies employed to control postharvest diseases today include (1) inoculum reduction, (2) prevention or eradication of field infections, (3) inactivation of wound infections; and (4) suppression of disease development and spread. Eckert and Ogawa[19,20] discussed how chemical control strategies are applied to many commodities.

Development of postharvest microbial biocontrol agents shares some methodology and regulatory aspects with those of chemical control, but additional issues and complexities arise. At the time of this writing, postharvest biocontrol agents have not been approved for commercial use, although patents have been issued or are pending for several; and pilot-scale tests are in progress or have been completed for some.[22,23] However, microbial biocontrol of insects has been used successfully to replace insecticides for some applications, and research in this area is copious.[24] Many aspects of screening strategies, safety, mode of biocontrol action, and commercialization of insect biocontrol agents are well studied.[25] Classified as "biorational pesticides" by the U. S. Environmental Protection Agency (EPA), commercial use of insect-controlling microbes began in the United States in 1948 when the first of many *Bacillus* spp. was approved.[26] Biorational pesticides can be introduced with fewer demands for toxicological and residue testing before registrations are granted than can synthetic pesticides. This lower cost makes them more suitable for small markets, such as the demand to control postharvest decay.

II. INFLUENCE OF ETIOLOGY ON BIOCONTROL STRATEGIES

A. TIME AND LOCATION OF INFECTION

Most workers reported phyllosphere and postharvest biocontrol of necrotrophic pathogens to be protectant in action, although a few reported minor eradicant action and none reported suppression of pathogen spread. Once infection takes place, control becomes more difficult.[7] A successful strategy requires detailed knowledge of the pathogen etiology to be able to exploit the protectant action of the antagonists. For an antagonist to best interrupt the disease cycle, it is critically important to know when, where, and how infection occurs so it is present before or at the time of infection. Most success with biocontrol agents has resulted from prevention of field infections[14,27,28] and protection of wounds from infection.[7,8,10,12,13] The time and location of infection by pathogens causing postharvest decay varies.[29] Infection can occur as early as bloom on flower parts, such as infections by *Botrytis cinerea* on grapes and strawberries, or long after harvest by pathogens growing between hosts in storage, such as the fast-spreading decays of berries and fruits caused by *Botrytis cinerea* and *Rhizopus* spp.[29] It can occur between flowering and fruit maturation, such as the stem end rots of citrus[30] and mangoes.[31] Infection can occur early in the crop cycle through senescent flower parts, through natural openings such as lenticels or stomates, or through mechanical wounds caused by insect feeding injuries, thorns, or hail.[29] Infections by wound pathogens occur when the wounds are made or soon afterward.[29]

B. PREHARVEST TREATMENTS
Studies evaluating preharvest treatments to control postharvest decay are uncommon, and this area deserves more attention. Although a postharvest treatment is usually more effective and efficient, a near-harvest treatment is an appropriate strategy in situations where numerous infections at harvest are expected or handling practices make postharvest treatments difficult to apply promptly. Preharvest strategies to control postharvest decay have used field sprays of fungicides,[19] and the limited biocontrol efforts employing preharvest applications have had some success.[18,27,28,32] For example, *Trichoderma* spp. applied to grapes or strawberries before harvest significantly reduced postharvest decay by *Botrytis cinerea*.[18,32] Many phyllosphere saprophytes reduce preharvest infections by necrotrophic pathogens such as *Botrytis*, *Alternaria*, and *Colletotrichum* by about 50%.[2,14] Since many postharvest pathogens perpetuate on crop debris in the field, a little-tested opportunity for control of these organisms is by antagonism of this phase of the life cycle. Applications of *Trichoderma* spp. late in the prior season decreased numbers of sclerotia and conidia of *Botrytis cinerea* in grape vines early the following spring.[18] Applications to grape flowers also were effective in reducing the number of infections on berries at harvest.[18]

C. WOUND PATHOGENS
Many pathogens responsible for postharvest diseases are unable to penetrate the host except through wounds. Mechanical and physiological injuries created during and after harvest are the usual infection courts used by these pathogens, and they cause the most devastating postharvest losses.[29] Wounds greatly facilitate infection even for pathogens that can penetrate the cuticle directly, such as *Monilinia fructicola* on stone fruit and *Botrytis cinerea* on many hosts. The severity of postharvest diseases induced by pathogens that exploit mechanical wounds is usually proportional to the number and severity of wounds inflicted during harvest.[19] The natural wound created by severing the crop from the plant and other natural openings such as lenticels and stomates also are frequent routes of infection. Nonsystemic protectants applied before infection do not arrest wound infections, since the pathogen exploits the injuries to access unprotected host tissue beneath the cuticle. Decay from wound infections is reduced by minimizing injuries, reducing inoculum density, or using eradicating treatments. Often, wounds become resistant to infection within days or even hours.[19] Presumably it is not essential that an antifungal treatment persist at an injured site.

D. INACTIVE INFECTIONS
Particularly problematic diseases for antagonists to control are those characterized by inactive infections. Pathogenesis ceases after infection due to the inherent or induced resistance of the host tissue, or to some morphological barrier. This inactive state of

infection, termed "latent" if not visible to the eye and "quiescent" if visible, persists until the resistence of the host declines with advancing maturity or senescence after harvest.[19] Examples of quiescent or latent infections include *Colletotrichum* infections on bananas, mangoes, papayas, avocados, and citrus fruit;[19] *Botrytis cinerea* infections on grapes and strawberries;[29] stem end rots of citrus[30] and mangoes;[31] *Monilinia fructicola* on stone fruits;[33] and lenticel rots of apples and potatoes.[19] These inactive infections are difficult to control. Protective fungicides and presumably antagonists provide some control if present before infection occurs. Systemic fungicides[19,20] and heat treatments[34] can eradicate infections after they occur.

III. POSTHARVEST BIOCONTROL SCREENING METHODS

A. ORIGIN OF CANDIDATE ANTAGONISTS

From which environment should microbes for screening be obtained? Most investigators use naturally occurring microorganisms from fruit or vegetable surfaces.[7,8,10,35-37] The natural occurrence of candidate antagonists on hosts may suggest that they would have a superior ability to colonize fruit or vegetable hosts efficiently. It is reasoned that they would be good candidates for biocontrol because of their adaptation to the host and its environment (i.e., nutrients, water status) and perhaps because antagonism against pathogens may already exist.[35] However, surface microfloras on the host at harvest are organisms that colonized and persisted during the relatively dry and warm production phase of the crop in the field. These may not be the best candidates for postharvest use on mature fruits or vegetables under cool, moist storage conditions. Janisiewicz[7,8] suggested emphasis on screening for organisms that control apple diseases be placed on those collected near harvest or later after several months of storage. Stretch[37] found more antagonists for blueberry and cranberry fruit rots among microbes isolated during cold periods of the year from these hosts than among isolations during warm periods.

Cook[38] suggested that antagonists should be sought in areas where no disease occurs even though there is considerable inoculum pressure. Phillips[38a] used similar reasoning in the selection of antagonists to protect wounds from infection by *Monilinia fructicola,* the cause of brown rot of stone fruit. He selected candidates from among wound colonizing microbes, rather than those washed from the fruit surface. Presumably the only organisms selected were those competent to colonize and protect wounds from infection. Phillips punctured hundreds of peaches and inoculated each wound with spores of *M. fructicola.* After several days incubation, all but a very few wounds had lesions. The wounds of those few that escaped infection were colonized with the yeasts *Exophilia mansonii* or *Aureobasidium prunicola.* Subsequent testing showed these yeasts had protectant activity against brown rot.[39]

Although the microflora of the host has been a rich source of antagonists, the superiority of this method has not been demonstrated and the selection of candidates by other reasoning cannot be excluded. Spurr[16] found no relationship between the origin of an antagonist and its effectiveness against foliar pathogens, and this situation is probably true for postharvest pathogens as well.[11] Andrews[14] suggested many organisms rare to plants could colonize the phyllosphere under favorable conditions. Research on introduced bacteria showed that the deciding factor for residency was the ability to survive unfavorable conditions, not whether the bacteria grew under favorable conditions.[40] Furthermore, isolation of unknown agents from the host microflora entails additional time and expertise to identify genus and species of the candidate organisms. Others have screened candidate antagonists from: (1) known commercial fermentation agents,[41] (2) microorganisms that exhibit plant disease control action in other than postharvest applications,[42,43] or (3) pathogens with debilitated virulence.[44] Phyllosphere biocontrol agents have been obtained from soil[17,45] and from among insect biocontrol agents.[17]

B. SCREENING ON MEDIA

In vitro screening employs co-culture of the pathogen and candidate biological control agent on agar or other media. Suppression of the pathogen's growth is used to select agents for further testing. *In vitro* screening has been used to successfully identify effective biocontrol agents, and its use has been recommended by some authors because of its convenience and the evidence it provides regarding mode of biocontrol action.[14,42] However, antibiosis in culture and disease suppression on the host often show little or no correlation.[46] Ferreira[47] found no correlation between inhibition of *Botrytis cinerea* by epiphytic bacteria on potato dextrose agar and grape berries. Furthermore, the significance of antibiosis observed *in vitro* regarding the mode of biocontrol action also must be interpreted with care, since the production of antibiotics *in vitro* and the ability to control disease *in vivo* may only be coincidental.[46]

C. SCREENING ON THE HOST

In vivo screening employs placement of the pathogen and candidate biological control agent on the host. Subsequent suppression of disease is used to select agents for further testing. *In vivo* screening is superior to *in vitro* screening because: (1) ability of the candidate to survive on the host is tested; (2) modes of action other than antibiosis are selected; and (3) other problems, such as virulence against the host, can be seen. Fruits and vegetables are relatively small, inexpensive experimental units. It is easy to closely simulate industry storage conditions. Tests are of short duration (1 to 3 weeks) with many commodities. Candidates that are likely to have genuine promise are identified in a single experiment. Fruits or vegetables used in these tests should be free of excessive pesticide residues. However, growth regulators or pesticides that are in common industry use can be present, since the intolerance of an antagonist to them poses additional barriers to its commercial use. Fruit should represent the maturity stages distributed commercially. Roberts[36] obtained superior control of gray mold and *Mucor* rot on less mature fruit and poor control on mature fruit with both apples and pears with all *Cryptococcus* strains tested. Control of citrus green mold on lemons by *Pseudomonas cepacia,* however, was not influenced by fruit maturity.[43]

Spurr[48] discussed general aspects and pitfalls of the use of *in vitro* bioassays for the evaluation of biocontrol agents. Janisiewicz[7,8] developed primary and secondary *in vivo* screening strategies based on artificial inoculation of pome fruit, and the reader is encouraged to consult this work. A hybrid system employing both *in vitro* and *in vivo* aspects was used by Peng and Sutton[49] to screen microbes for the control of *Botrytis cinerea.* Strawberry leaf disks were spray-inoculated with pathogen spores and candidate microbes, and the extent of leaf colonization by the pathogen was estimated. Later suppressive isolates were applied to plants, and the incidence of fruit rot was determined at harvest. Several antagonists found by this method were as effective as captan or chlorothalonil. Similarly, Rhodes et al.[50] developed a preliminary screening method using tuber slices to find biocontrol antagonists to control bacterial tuber rots of potatoes.

D. ARTIFICIAL INOCULATION METHODS

Candidate antagonists should be applied to fruit under conditions which simulate those in which commercial losses occur. Since natural infection can be sparse or irregular, some artificial system for inoculation must often be developed so that extremely large numbers of fruits or vegetables need not be used. The inoculum should be applied in the proper infection courts, at the proper time, and in densities typical of commercial conditions. The selection of the inoculation method used for *in vivo* screening is critical to a successful strategy, since the selection of a screening system that does not simulate natural inoculation may identify antagonists as promising when they will have little value in commercial situations; while useful antagonists may escape notice. Janisiewicz[7] found infections originating from smooth cut wounds were easier to control than those from nail punctures; the nail wound more closely simulates wounds actually made during commercial pear operations than the smooth wound. Pusey et al. [51] assessed peach brown rot and Rhizopus rot control by *Bacillus subtilis* using naturally inoculated fruit and fruit artificially inoculated with *Monilinia fructicola* and *Rhizopus stolonifer* spores into either wounds or uninjured fruit surfaces. The decay of artificially inoculated fruit was reliably controlled, while irregular control of decay was observed with naturally inoculated fruit.

The inoculum density used should produce an intermediate frequency (50 to 80%) of decayed fruits or vegetables among control treatments. This optimizes the resolution of statistical mean separation procedures among treatments,[52] and it is likely to be a better simulation of disease under commercial conditions. De Matos[42] conducted preliminary experiments with each collection of lemons he obtained to determine the spore inoculum density of *Penicillium digitatum* required to obtain 50 to 80% decay. Then he conducted tests that evaluated antagonist effectiveness at this inoculum density. Alternatively, ranges of both pathogen inoculum and antagonist propagule densities can be employed to thoroughly characterize the potency of an antagonist.[8] Negative (uninoculated) controls should be employed to determine whether significant natural inoculation has interfered with an artificial inoculation screening test. It is recommended that only one inoculation per fruit is used, since wounding or infection from one site may interfere with disease development at other inoculation sites on the same host.[29] A fungicide used after harvest can be included as a separate treatment to facilitate comparison of the magnitude of control of decay by candidate antagonists.

E. BIOCONTROL AGENT APPLICATION

Microbes can be grown on semisolid or liquid media. Liquid media are superior because the distribution of nutrients is more uniform, resulting in predictable and more synchronous growth than that on semisolid media.[53] To maximize the yield of viable cells, cultures should be used no later than the late exponential growth stage. Batch cultures can be used in preliminary tests. For large-scale tests, production of large volumes is facilitated by the use of a fermentor to continuously supply fresh nutrients so that exponential growth is maintained indefinitely. The propagules of the microbe can be applied in the culture broth or centrifuged free of this medium and suspended in water or buffer for use. The culture fluid in which the microbe was cultured, filtered to remove microbe cells, can be applied as another treatment since it may contain antifungal compounds. Antagonist cells can be applied with nutrients to determine the influence of nutrient availability on disease suppression. These additional treatments have been used to provide some evidence regarding the mode of biocontrol action.[36]

Fruits or vegetables can be dipped or sprayed with candidate antagonists, or pressure infilitration can be used to facilitate penetration of the antagonist beneath the cuticle or into stem tissue.[54] Spray application is more convenient because smaller quantities of the culture are required, and contamination of the culture with other organisms during repeated dipping of fruit is avoided. Methods of spray application of pesticides — including control of droplet size, nozzle design, and other subjects — were described by Matthews.[55] Nitrogen gas or air as propellants are preferable over carbon dioxide, since carbon dioxide under pressure inhibits some microbes.[56] In pilot-scale tests, waxes, foams, fungicides, or other adjuvants also may be added.

Most workers elect to apply antagonists before or at the time of inoculation. For example, *Rhodotorula* sp.,[42] *Trichoderma viride*,[42] *Debaryomyces hansenii*,[35,57] and *Aureobasidium pullulans*[57] effectively controlled citrus green mold in this type of application. Others applied propagules of the pathogen first, followed later by an antagonist. This latter protocol identifies antagonists with eradicant action. Eradicant action is useful for the control of wound-invading pathogens that inoculate wounds about the same time the wound is made, such as the *Penicillium* blue and green molds on citrus fruit.[30] These pathogens enter through fresh wounds inflicted during the harvest and postharvest handling operations. Therefore, an agent that can eradicate the nascent wound infections, generally from 6 to 24 h old, is required. The fungicide imazalil controlled decay effectively if applied 24 h after inoculation, while *Pseudomonas cepacia* was effective only if applied within 12 h after inoculation.[43] The yeast *Debaryomyces hansenii* was effective if applied at 3 h but not at 7 h after inoculation.[35]

F. POSTTREATMENT INCUBATION AND EVALUATION

Incubation conditions should simulate those encountered in industry practice. A comprehensive description of industry practices for all major horticultural crops was recently issued.[58] Simulation of both the storage duration and environment is important. Storage longer than required for disease development among control treatments is recommended, because antagonists may only delay the onset of symptoms, rather than provide persistent control;[43,59] or the antagonist itself may injure the fruit.[43] Although optimum temperature and humidity storage conditions have long been known for most fruits and vegetables,[60] actual storage and transportation conditions are often warmer than those recommended[61-64] and may influence biocontrol efficacy. Gullino et al.[65] reported *Trichoderma* spp. suppression of gray mold on grapes was improved by high humidity and increasing the temperature from 15 to 28°C.

Disease can be quantified by incidence, severity, or both. Incidence is usually recorded as a percentage of infected fruit or inoculation sites, while severity is generally recorded as lesion size. Fungicide evaluations often include some measure of sporulation intensity.[30] The selection of these parameters should be based on the type of losses that occur under commercial conditions.

The most effective antagonists deserve special attention. To conduct pilot-scale tests, collaboration with knowledgeable industry cooperators is recommended. Since a patent can be crucial to commercialization and marketing of biocontrol antagonists, careful consideration should be given to applying for patents prior to releasing information that could make an agent unpatentable.[67,68] However, patents will only provide protection for a specific strain of a specific organism and not preclude others from commercializing similar antagonists later.

G. FATE OF APPLIED BIOCONTROL ANTAGONISTS

Spurr and Knudsen[17] recommended that whenever biocontrol agents are applied to plants their populations should be monitored at several reasonably close intervals. In wounds on fruit, effective postharvest biocontrol antagonists grow rapidly and are persistent.[36,39,43,66] Conversely, populations of phyllosphere-applied antagonists increase little or decline.[14,15,17] The moist, nutrient-rich wounds and the conducive postharvest environment compared to the relatively harsh phyllosphere environment are factors responsible for this difference. Employing antagonists with antibiotic- or fungicide-resistance or molecular markers facilitates studies of their population dynamics, especially if interference from contaminating organisms is common. These strategies have been employed for foliar[15] and soil antagonists.[38]

H. NUTRIENTS

Most postharvest pathogens are unspecialized necrotrophs that grow saprophytically in many environments, often taking up nutrients that later facilitate pathogenesis. Nutritional amendments could increase the efficacy of antagonists by providing and possibly creating ecological niches that would be favorable to the antagonist or deleterious to the pathogen. Nutrient composition and quantity influenced populations of pathogenic and saprophytic microorganisms on aerial plant surfaces.[69-71] Janisiewicz et al.[66] evaluated carbon and nitrogen sources to enhance the antagonism of *Pseudomonas syringae* against *Penicillium expansum*, the cause of postharvest blue mold of apples. The amino acids L-asparagine and L-proline greatly enhanced efficacy in artificially inoculated wounds. Gullino et al.[65] improved suppression of *Botrytis cinerea* in detached grape clusters by the addition of 1 g of malt per liter to the inoculum of antagonistic *Trichoderma* spp. Postharvest calcium treatments to apples not only increased resistance to pathogens and improved storage quality,[13] but also enhanced the antagonist activity of some yeasts to pathogens on apples.[73,74] Calcium also enhanced biocontrol of *Botrytis cinerea* on grapes.[65]

Organic amendments also have enhanced the efficacy of foliar antagonists. The survival and efficacy of the chitinolytic bacterial antagonist, *Bacillus cereus*, was enhanced by chitin applications to peanut leaves.[75] Hydrolyzed cellulose and a vegetable oil-based spreader-sticker enhanced the foliar tenacity and control of several apple diseases by the antagonist *Chaetomium globosum*.[76] Carboxymethyl cellulose improved adhesion and supported conidial germination of the antagonist *Trichoderma harzianum* on apples.[28] Carboxymethyl cellulose improved control of *Botrytis cinerea* when applied with the antagonist, although it increased disease when applied alone.[18]

I. INFLUENCE OF OTHER TREATMENTS

The influence of pesticides on biocontrol antagonists has been investigated. Tronsmo[77] found 18 of 21 fungicides, and all 12 insecticides tested to be strongly inhibitory *in vitro* to the antagonists *Trichoderma viride* and *T. harzianum*. These pesticides could interfere with biocontrol by these antagonists and should be avoided. Conversely, enhanced disease control was observed when fungicides were combined with mutant or natural fungicide-resistant antagonists. Significantly improved disease control was obtained when fungicides were combined with antagonists applied to apples,[28] grapes,[32,65] peaches,[23,51] strawberries,[27] and citrus.[22] In addition to application with fungicides, biocontrol antagonists could be applied in mixtures with other antagonists to control multiple diseases simultaneously when no single broad spectrum antagonist is available. Antagonists should integrate into commercial practices with other postharvest treatments, such as waxes, antiscald compounds, controlled atmospheres, and growth regulators.

Integration with other alternative decay control technologies has been little investigated. Combination on postharvest biocontrol with strategies that do not have persistent protectant residues,[1] such as hot water, high temperature curing, sodium hypochlorite, ozone, or chlorine dioxide, could benefit from the persistent protection conferred by biocontrol antagonists. While heat treatment eradicates latent infections and greatly reduces surface microflora populations, it does not provide persistent protection of the commodity from reinfestation and often renders the crop more susceptible to infection.[34] High temperature conditioning or curing[29,30] to reduce decay employs temperatures that are optimal to the growth of many microbes, thereby greatly increasing the number of potential antagonists.

J. MOLECULAR APPROACHES TO ENHANCE BIOCONTROL

Lindemann[78] discussed approaches using genetic manipulation of biocontrol agents to enhance their efficacy. Baker[79] suggested approaching this objective in two phases: first, define and analyze biocontrol mechanisms; second, devise creative strategies for their enhancement. The introduction by molecular means of improved competitive ability (for nutrients or attachment), chemical resistance, ability to produce an antifungal antibiotic of known safety, or alteration of temperature optima of antagonists to improve safety or efficacy are all technically feasible strategies. Obtaining approval for engineered microbes has been more difficult than for native microbes.[80] Conditional registration for two genetically engineered microbial insect biocontrol agents was granted by the EPA in 1991. They contain *Pseudomonas fluorescens* strains that produce insecticidal endotoxins originally from *Bacillus thuringensis*. However, the cells of *P. fluorescens* are not viable in the commercial formulation.

K. SAFETY CONCERNS

Decisions about the selection of the candidate antagonists, especially if they are obtained from culture collections, should be influenced by their safety. A review of safety issues of postharvest biocontrol antagonists is addressed elsewhere in this volume. Currently, safety is an area of energetic debate[13,38] and of limited regulatory precedent.[26] Selecting antagonists of established safety would facilitate obtaining regulatory approval. Rogoff[81] described three approaches to evaluate human safety concerns related to biorational pesticides. These were (1) "infectivity" as a parasite, evidenced by persistence, replication, or colonization of nontarget organisms; (2) "virulence-toxicity" implying potential for direct injury to nontarget organisms of an acute, subacute, or chronic nature; and (3) "hypersensitivity" implying a potential for a deleterious immune response. Unlike the biorational pesticides currently approved, postharvest biocontrol involves treatment of the product the consumer ingests with high propagule numbers of viable and persistant antagonists.

No virulent microbial pathogen of man deserves evaluation for disease control because of its obvious risk. Brackett[82] listed human pathogens of concern on fresh produce. They occur primarily as a consequence of contamination and are not normal residents.[83] The opportunistic human virulence of many other microbes that may be residents on fruits and vegetables is a potential safety issue. Although healthy people are immune from infection from all except human pathogens, many other organisms are opportunistic colonizers of people who are immunosuppressed, such as organ transplant recipients, or have other underlying conditions, such as burn victims or people under prolonged antimicrobial therapy.[84] It has been suggested that the natural occurrence of some antagonists on fruits and vegetables should relieve concerns about their safety, since they have passed a long test of time without apparent harm to consumers.[35] However, microbes isolated from fresh produce — including *Serratia marcescens, Klebsiella* spp., and *Pseudomonas* spp. — have been associated with clinical human infection.[83,85] Stoddart[86] stated some investigators consider hospital food a major source of hospital infections. Fresh produce and ornamental plants in hospital environments increase the risk of introduction of these microbes and infection of debilitated patients.[87,88] In spite of these issues, genera of these opportunistic colonizers have had registrations for some uses granted. For example, *Pseudomonas fluorescens* EG-1053 was subjected to a variety of toxicity and virulence testing; it was determined to be safe and registration was granted in 1988 by the EPA for its use as a soil fungicide for cotton. Concerns about mammalian virulence are alleviated by the selection of biocontrol antagonists that have never appeared in clinical isolations[84] or those that cannot grow at human body temperature (37°C). Since the optimum storage temperatures for most fresh fruits and vegetables are 15°C or less,[60] this should be achievable.

The production of antibiotics or secondary metabolites is a safety concern that can influence the selection of candidate organisms, because these metabolites conceivably could be of environmental or animal toxicological consequence. Biocontrol action dependent on antibiotics comprises an additional regulatory hurtle, since these "microbial fungicides" do not differ in concept from the conventional chemicals that were developed for their replacment. There is extensive literature on safety and the reasoning used in the regulation of toxin-producing bacteria to control insects, and the reader is encouraged to review this work.[24] Antifungal antibiotic-producing bacteria on plants are common. Of more than 10,000 bacteria isolated from roots on grapevines and other crops, more than one third showed *in vitro* antibiosis to the fungus *Colletotrichum lindemuthianum*.[89] This is true on fruit as well. For example, total microbe populations on grapefruit are about

1 x 10^5 colony-forming units per square centimeter,[90] and antibiotic-producing microorganisms commonly occur among those characterized from lemons.[42] Antifungal antibiotics produced by *Bacillus* and *Pseudomonas* spp. isolates are good postharvest fungicides.[91,92] Pyrrolnitrin, produced by *P. cepacia,* has a low order of mammalian toxicity, and no toxicological hazards have been identified.[93] However, other antibiotics can be of concern. For example, *Myrothecium* spp. controlled postharvest green mold of citrus,[94] but the severe toxicological hazard of the tricothecene mycotoxins produced by this genus[95] precludes further exploitation of this beneficial phenomena.

The issue of introducing antibiotics to food in order to control spoilage has been addressed.[56] Ideally, the antibiotic should decompose into innocuous products or be destroyed on cooking or by digestive enzymes after ingestion. It should not be used in foods if the same compound is used as a therapeutic agent or as an animal feed additive. Two antibiotics (nisin and natamycin) are approved in many countries for use in food, and many other promising antibiotics have been investigated.[56] Lactic acid bacteria occur in many dairy and fruit juice products as fermentation or spoilage antagonists, and present no safety hazard in respect to human virulence or secondary metabolite production.[56] They owe their dominance in these products at least partially to the production of antibiotics,[96] some of which are antifungal.[97] Regulators probably will demand assessment of the safety of an antibiotic if it is introduced in the diet of consumers at significant concentrations.

L. MODE OF BIOCONTROL ACTION

This subject is beyond the scope of this review, and the reader is encouraged to look elsewhere in this volume for reviews on this subject. Most workers recommended biocontrol action based on induced resistance in the host or competition for space or nutrients, rather than antibiosis, for reasons of human safety. Janisiewicz[8] suggests isolates also be tested for *in vitro* antibiosis during screening, since this is an important first step in compiling evidence concerning safety and mode of biocontrol action. The role of antibiotic production in plant disease biocontrol has been investigated by molecular techniques.[98] A mode of action role for an antibiotic is supported if disease suppressiveness positively correlated with antibiotic production. Similarly, the role of antibiotic production can be investigated by the acquisition of high levels of antibiotic resistance by the pathogen. After the acquisition of a 600-fold resistance to pyrrolnitrin by isolates of the citrus pathogen *Penicillium digitatum,* they were still readily controlled by a pyrrolnitrin-producing strain of *Pseudomonas cepacia,* suggesting biocontrol was not provided solely by the action of this antibiotic.[43]

IV. A CASE STUDY: BIOCONTROL OF BROWN ROT
OF STONE FRUITS

An *in vivo* screening protocol developed in my laboratory was as follows: each fruit was punctured once with a 3 x 3 mm tool; this site was inoculated with about 200 conidia of *Monilinia fructicola;* and 2 h later cells or spores of candidate antagonists were applied. After incubation for several days, the number and size of decay lesions were recorded. About 90% of the inoculated control fruit had decay lesions within 4 d. Superior activity by *Pseudomonas corrugata* was identified after screening many antagonists. The bacterium reduced decay 90% or more and was compatible with commercial waxes. It multiplied rapidly in wounds. The mode of action was undetermined. Although it produced clear zones when cocultured with *M. fructicola* on agar, washed cells of the antagonist controlled decay while the cell-free culture fluid did not. *P. corrugata* was later applied to peaches and nectarines after harvest under commercial conditions. These fruit had a very high decay potential; untreated fruit decayed from 55 to 97% in several days at 20°C. Viable spores of *M. fructicola,* determined on a selective medium,[99] ranged from

20,000 to 500,000 per fruit. Fruit were dipped 1 min in dense (1 x 10^9 colony-forming units per milliliter) cultures of *P. corrugata* or 600 μg/ml imazalil, a fungicide registered for use on stone fruit. In repeated tests, *P. corrugata* provided little or no reduction in decay, while the fungicide reduced decay 90% or more.

The screening process did not simulate this disease as it occurs under commercial conditions. To develop a revised strategy, the etiology of the pathogen must be scrutinized closely. The modes of infection by which *M. fructicola* initiates postharvest decay are diverse[100-102] and complicate the selection of a screening strategy. The infection processes have been described, but the contribution of the various types of infection to the total decay losses has not been rigorously quantified and probably varies with each season and location. The initial inoculum is composed of conidia produced from the sporulation of mummified fruit of the prior year, lesions on twigs, peduncles, and blossoms.[103,104] By midseason, lesions on new fruit produce copious inoculum.[67] Conidia survive dry periods, even after germination.[105] These conditions kill most vegetative bacteria. The number of fruit infections is influenced by the quantity and quality of conidial inoculum on the fruit,[67,99,106,107] the activity of insects and birds as agents of injury and spore dispersal,[102,108] the duration of conducive moisture and temperatures during flowering and when fruit approaches harvest,[100] and the number of mechanical injuries inflicted during harvest and packing.[102,109] *M. fructicola* infections originate from: (1) preharvest quiescent or latent infections and (2) postharvest conidial germination and penetration on mature fruit.[103,109] Quiescent and latent infections occur in fruit at various stages of immaturity, and decay is delayed until the fruit matures and ripens.[33,101,106] Conversely, conidia can also germinate, penetrate the fruit, and cause visible lesions in 24 to 48 h, depending on whether the infection court is a wound or the intact surface of the fruit.[100,106]

The etiology of this pathogen and prior biocontrol results suggest an integrated approach to brown rot control, using preharvest and postharvest treatments, could be attempted to attain acceptable levels of control. Biocontrol antagonists applied before harvest may reduce the production of inoculum, reduce the number of latent infections, or colonize preharvest injuries. De Cal and co-workers[110] reduced the incidence of peach twig blight, caused by the closely related fungus *Monilinia laxa,* by 38 to 80% with applications of *Penicillium frequentans* in field trials. The conidia of the antagonist were suspended in nutrients (malt and yeast extracts) that improved colonization and persistence of *P. frequentans* on the shoots. Because insect feeding injures and introduces inoculum into the fruit flesh, control of them is important.[108] After harvest, biocontrol antagonists of choice would be those with potent antibiotic action that control decay initiating on the fruit surface or from within wounds, such as provided by *Bacillus subtilis.*[41,51] Alternatively, protectant action by antagonists might prove beneficial when applied to protect fruit after hot water treatment.[34]

REFERENCES

1. **Pusey, P. L., Wilson, C. L., and Wisniewski, M. E.,** Management of postharvest diseases of fruits and vegetables: strategies to replace vanishing fungicides, in *Pesticide Interactions in Crop Production: Beneficial, Deleterious Effects,* Altman, J., Ed., CRC Press, Boca Raton, FL, 1993.
2. **Ames, B. N. and Gold, L. S.,** Too many rodent carcinogens: mitogenesis increases mutagenesis, *Science,* 249, 970, 1990.
3. **Gold, L. S., Slone, T. H., Stern, B. R., Manley, N. B., and Ames, B. N.,** Rodent carcinogens: setting priorities, *Science,* 258, 261, 1992.
4. **Archibald, S. O. and Winter, C. K.,** Pesticides in our food. Assessing the risks, in *Chemicals in the Human Food Chain,* Winter, C. K., Seiber, J. N., and Nuckton, C. F., Eds., Van Nostrand Reinhold, New York, 1990, 1.

5. **Scheuplein, R. J.,** Perspectives on toxicological risk — an example: foodborne carcinogenic risk, *Crit. Rev. Food Sci. Nutr.*, 32, 105, 1992.
6. National Research Council, *Pesticide Resistance. Strategies and Tactics for Management,* National Academy Press, Washington, D.C., 1986.
7. **Janisiewicz, W. J.,** Control of postharvest diseases of fruits with biocontrol agents, in *The Biological Control of Plant Diseases,* Book Series No. 42, Food and Fertilizer Technology Center, Taipei, Taiwan, 1991, 56.
8. **Janisiewicz, W. J.,** Biological control of diseases of fruits, in *Biocontrol of Plant Diseases,* Mukerji, K. G. and Garg, K. L., Eds., CRC Press, Boca Raton, FL, 1988, 153.
9. **Pusey, P. L.,** Control of pathogens on aerial plant surfaces with antagonistic microorganisms, *Biol. Cult. Tests Con. Plant Dis.*, 5, v, 1990.
10. **Moline, H. E.,** 1991. Biocontrol of postharvest bacterial disease of fruits and vegetables, in *Biological Control of Postharvest Diseases of Fruits and Vegetables,* Agricultural Research Service, U.S. Department of Agriculture, ARS-92, 1991, 114.
11. **Wilson, C. L. and Wisniewski, M. E.,** Biological control of postharvest diseases of fruits and vegetables: an emerging technology, *Annu. Rev. Phytopathol.*, 27, 425, 1989.
12. **Wilson, C. L., Wisniewski, M. E., Biles, C. L., McLaughlin, R., Chalutz, E., and Droby, S.,** Biological control of post-harvest diseases of fruits and vegetables: alternatives to synthetic fungicides, *Crop Prot.*, 10, 172, 1991.
13. **Wisniewski, M. E. and Wilson, C. L.,** Biological control of postharvest diseases of fruits and vegetables: recent advances, *HortScience*, 27, 94, 1992.
14. **Andrews, J. H.,** Strategies for selecting antagonistic microorganisms from the phylloplane, in *Biological Control on the Phyllosphere,* Windels, C. E. and Lindow, S. E., Eds., American Phytopathological Society, St. Paul, MN, 1985, 31.
15. **Andrews, J. H.,** Biological control in the phyllosphere, *Annu. Rev. Phytopathol.*, 30, 603, 1992.
16. **Spurr, H. W., Jr.,** Experiments on foliar disease control using bacterial antagonists, in *Microbial Ecology of the Phylloplane,* Blakeman, J. P., Ed., Academic Press, New York, 1981.
17. **Spurr, H. W., Jr. and Knudsen, G. R.,** Biological control of leaf diseases with bacteria, in *Biological Control on the Phylloplane,* Windels, C. E. and Lindow, S. E., Eds., APS Press, St. Paul, MN, 1985, 45.
18. **Tronsmo, A.,** Use of *Trichoderma* spp. in biological control of necrotrophic pathogens, in *Microbiology of the Phyllosphere,* Fokkema, N. J. and van den Huevel, J., Eds., Cambridge Press, London, 1986, 348.
19. **Eckert, J. W. and Ogawa, J. M.,** The chemical control of postharvest diseases: subtropical and tropical fruits, *Annu. Rev. Phytopathol.*, 23, 421, 1985.
20. **Eckert, J. W. and Ogawa, J. M.,** The chemical control of postharvest diseases: deciduous fruits, berries, vegetables and root/tuber crops, *Annu. Rev. Phytopathol.*, 26, 433, 1988.
21. **Davidse, L. C.,** Benzimidazole fungicides: mechanism of action and biological impact, *Annu. Rev. Phytopathol.*, 24, 43, 1986.
22. **Hofstein, R., Droby, S., Chalutz, E., Wilson, C., and Fridlender, B.,** Scaling-up the production for application of an antagonist — from basic research to R & D, in *Biological Control of Postharvest Diseases of Fruits and Vegetables,* Agricultural Research Service, U.S. Department of Agriculture, ARS-92, 1991, 197.
23. **Pusey, P. L., Hotchkiss, M. W., Dulmage, H. T., Baumgardner, R. A., Zehr, E. I., Reilly, C. C., and Wilson, C. L.,** Pilot scale tests for commercial production and application of *Bacillus subtilis* (B-3) for postharvest control of peach brown rot, *Plant Dis.*, 72, 622, 1988.

24. **Kurstak, E., Ed.,** *Microbial and Viral Pesticides,* Marcel Dekker, New York, 1982.

25. **Ferguson, M. P., and Alford, H. G.,** Microbial/biorational pesticide registration, Special publication 3318, University of California, 1986.

26. **Mendelsohn, M., Rispin, A., and Hutton, P.,** Environmental Protection Agency oversight of microbial pesticides, in *Biological Control of Postharvest Diseases of Fruits and Vegetables,* Agricultural Research Service, U.S. Department of Agriculture, ARS-92, 1991, 234.

27. **Gullino, M. L., Aloi, C., and Garibaldi, A.,** Chemical and biological control of grey mould of strawberry, *Med. Fac. Landbouww. Rijksuniv. Gent.,* 55, 967, 1990.

28. **Tronsmo, A.,** Biological and integrated controls of *Botrytis cinerea* on apple with *Trichoderma harzianum, Biol. Control,* 1, 59, 1991.

29. **Snowdon, A. L.,** *A Color Atlas of Post-Harvest Diseases and Disorders of Fruits and Vegetables,* Vol. 1 and 2, CRC Press, Boca Raton, FL, 1990.

30. **Eckert, J. W. and Eaks, I. L.,** Postharvest disorders and diseases of citrus fruits, in *The Citrus Industry,* Vol. 4, Reuther, W., Calavan, E. C., and Carman, G. E., Eds., University of California Press, Berkeley, 1989, 179.

31. **Johnson, G. I., Mead, A. J., Cooke, A. W., and Dean, J. R.,** Mango stem end rot pathogens — fruit infection by endophytic colonization of the inflorescence and pedicel, *Ann. Appl. Biol.,* 120, 225, 1992.

32. **Elad, Y. and Zimand, G.,** Experience in integrated chemical-biological control of grey mould (*Botrytis cinerea*), *Bull. SROP,* 14, 195, 1991.

33. **Kable, P. F.,** Significance of short-term latent infections in the control of brown rot in peach fruits, *Phytopathol. Z.,* 70, 173, 1971.

34. **Couey, H. M.,** Heat treatment for control of postharvest diseases and insect pests of fruits, *HortScience,* 24, 198, 1989.

35. **Chalutz, E. and Wilson, C. L.,** Postharvest biocontrol of green and blue mold and sour rot of citrus fruit by *Debaryomyces hansenii, Plant Dis.,* 74, 134, 1990.

36. **Roberts, R. G.,** Characterization of postharvest biological control of deciduous fruit diseases by *Cryptococcus* spp., in *Biological Control of Postharvest Diseases of Fruits and Vegetables,* Agricultural Research Service, U.S. Department of Agriculture, ARS-92, 1991, 37.

37. **Stretch, A. W.,** Biological control of blueberry and cranberry fruit rots (*Vaccinium corymbosum* L. and *Vaccinium macrocarpon* Ait.), *Acta Hortic.,* 241, 301, 1989.

38. **Cook, R. J.,** Biological control of plant diseasses: broad concepts and applications, in *The Biological Control of Plant Diseases,* Book Series No. 42, Food Fertilizer Technology Center, Taipei, Taiwan, 1991, 1.

38a. **Phillips, D. J.,** personal communication, 1991.

39. **Smilanick, J. L., Denis-Arrue, R., Bosch, J. R., Gonzalez, A. R., Henson, D. J., and Janisiewicz, W. J.,** Biocontrol of postharvest brown rot of nectarines and peaches by *Pseudomonas* species, *Crop Prot.,* 12, 513, 1993.

40. **O'Brien, R. D. and Lindow, S. E.,** Effect of plant species and environmental conditions on epiphytic population sizes of *Pseudomonas syringae* and other bacteria, *Phytopathology,* 79, 619, 1989.

41. **Pusey, P. L.,** Antibiosis as mode of action in postharvest biological control, in *Biological Control of Postharvest Diseases of Fruits and Vegetables,* Agricultural Research Service, U.S. Department of Agriculture, ARS-92, 1991, 127.

42. **De Matos, A. P.,** Chemical and Microbiological Factors Influencing the Infection of Lemons by *Geotrichum candidum* and *Penicillium digitatum,* Ph.D. dissertation, University of California, Riverside, 1983.

43. **Smilanick, J. L. and Denis-Arrue, R.,** Control of green mold of lemons with *Pseudomonas* species, *Plant Dis.,* 76, 481, 1992.

38

44. **Lim, T.-K. and Rohrbach, K. G.,** Role of *Penicillium funiculosum* strains in the development of pineapple fruit diseases, *Phytopathology,* 70, 663, 1980.
45. **Levy, E., Eyal, Z., and Chet, I.,** Suppression of *Septoria tritici* blotch and leaf rust on wheat seedling leaves by pseudomonads, *Plant Pathol.,* 37, 551, 1988.
46. **Fravel, D. R.,** Role of antibiosis in the biocontrol of plant diseases, *Annu. Rev. Phytopathol.,* 26, 75, 1988.
47. **Ferreira, J. H. S.,** *In vitro* evaluation of epiphytic bacteria from table grapes for the suppression of *Botrytis cinerea, S. Afr. J. Enol. Vitic.,* 11, 38, 1990.
48. **Spurr, H. W., Jr.,** Bioassays — critical to biocontrol of plant disease, *J. Agric. Entomol.,* 2, 117, 1985.
49. **Peng, G. and Sutton, J. C.,** Biological methods to control grey mould of strawberry, *Brighton Crop Prot. Conf.,* 1, 233, 1990.
50. **Rhodes, D., Logan, C., and Gross, D.,** Selection of *Pseudomonas* spp. inhibitory to potato seed tuber decay, *Phytopathology,* 76, 1078, 1986.
51. **Pusey, P. L., Wilson, C. L., Hotchkiss, M. W., and Franklin, J. D.,** Compatibility of *Bacillis subtilis* for postharvest control of peach brown rot with commercial fruit waxes, dicloran, and cold-stroage conditions, *Plant Dis.,* 70, 587, 1986.
52. **Sokal, R. R. and Rohlf, F. J.,** *Biometry. The Principles and Practices of Statistics in Biological Research,* 2nd ed., W. H. Freeman, New York, 1981.
53. **Garraway, M. O. and Evans, R. C.,** *Fungal Nutrition and Physiology,* Wiley-Interscience, New York, 1984.
54. **Spalding, D. H.,** Evaluation of various treatments for control of postharvest decay of Florida mangos, *Proc. Fla. State Hortic. Soc.,* 99, 97, 1986.
55. **Matthews, G. A.,** *Pesticide Application Methods,* Longman, New York, 1979.
56. **Jay, J. M.,** *Modern Food Microbiology,* 4th ed., Van Nostrand Reinhold, New York, 1991.
57. **Wilson, C. L. and Chalutz, E.,** Postharvest biological control of *Penicillium* rots of citrus with antagonistic yeasts and bacteria, *Sci. Hortic. Canterbury Engl.,* 40, 105, 1989.
58. **Kader, A. A.,** *Postharvest Technology of Horticultural Crops,* University of California Press, Pub. No. 3311, 1992.
59. **Falconi, C. and Mendgen, K.,** Inhibition of postharvest pathogens by epiphytic bacteria isolated from *Malus comunis* cultivar Golden Delicious, in *Communications from the Federal Biological Institute for Agriculture and Forestry,* Berlin-Damlen, No. 266, 47th German Plant Protection Convention, Berlin, Germany, 1991, 336.
60. **Hardenburg, R. E., Watada, A. E., and Wang, C. Y.,** *The Commercial Storage of Fruits, Vegetables, and Florist and Nursery Stocks,* Agricultural Research Service, U.S. Department of Agriculture, Handbook No. 66, 1986.
61. **Harvey, J. M., Harris, C. M., Tietjen, W. J., and Serio, T.,** *Quality Maintenance in Truck Shipments of California Strawberries,* U.S. Department of Agriculture, Advances in Agricultural Technology Western Ser. No. 12, 1980.
62. **Harvey, J. M., Harris, C. M., and Rasmussen, G. K.,** Protective storage and transit environments for kiwifruit (*Actinidia chinensis* Planch.) and their effect on fruit quality, *Crop Prot.,* 5, 277, 1986.
63. **Hinsch, R. T., Rij, R. E., and Kasmire, R. F.,** *Transit Temperatures of California Iceberg Lettuce Shipped by Truck during the Hot Summer Months,* U.S. Department of Agriculture, Agricultural Research Service, Marketing Research, Report No. 1117, 1981.
64. **Hinsch, R. T. and Harris, C. M.,** Exporting fumigated nectarines by van container, *Rev. Int. Froid,* 15, 59, 1992.
65. **Gullino, M. L., Aloi, C., and Garibaldi, A.,** Evaluation of the influence of different temperatures, relative humidities and nutritional supports on the antagonistic activity of *Trichoderma* spp. against grey mold of grape, in *Influence of Environmental Factors on the Control of Grapes Pests, Diseases, and Weeds, Proc. EC Expert's Group,* Cavalloro, R., Ed., Instituto di Patologia Vegetale dell Universita, Torino, Italy, 1989, 231.

66. **Janisiewicz, W. J., Usall, J., and Bors, B.,** Nutritional enhancement of biocontrol of blue mold on apples, *Phytopathology,* 82, 1364, 1992.

67. **Phillips, D. J. and Harris, C. M.,** *Postharvest Brown Rot of Peaches and Inoculum Density of Monilinia fructicola* (Wint.), Agricultural Research Service, U.S. Department of Agriculture, SEA W-9, 1979.

68. **Wieder, S. C.,** Patent protection for microorganisms and their use in biocontrol, in *Biological Control of Postharvest Diseases of Fruits and Vegetables,* Agricultural Research Service, U.S. Department of Agriculture, ARS-92, 1991, 229.

69. **Blakeman, J. P. and Brodie, I. D. S.,** Inhibition of pathogens by epiphytic bacteria on aerial plant surfaces, in *Microbiology of Aerial Plant Surfaces,* Dickinson, C. H. and Preece, T. F., Eds., Academic Press, London, 1976, 529.

70. **Fokkema, N. J.,** Competition for endogenous for endogenous and exogenous nutrients between *Sporobolomyces roseus* and *Cochliobolus sativus, Can. J. Bot.,* 62, 2463, 1984.

71. **Morris, C. E. and Rouse, D. I.,** Role of nutrients in regulating epiphytic bacterial population, in *Biological Control on the Phylloplane,* Windels, C. E. and Lindow, S. E., Eds., APS Press, St. Paul, MN, 1985, 63.

72. **Conway, W. S., Sams, C. E., McGuire, R. G., and Kelman, A.,** Calcium treatment of apples and potatoes to reduce postharvest decay, *Plant Dis.,* 76, 329, 1992.

73. **McLaughlin, R. J., Wisniewski, M. E., Wilson, C. L., and Chalutz, E.,** Effects of inoculum concentration and salt solutions on biological control of postharvest diseases of apple with *Candida* sp., *Phytopathology,* 80, 456, 1990.

74. **Lodovica-Gullino, M., Aloi, C., Palitto, M., and Benzi, D.,** Attempts at biological control of postharvest diseases of apple, *Phytoparasitica,* 19, 258, 1991.

75. **Kokalis-Burelle, N., Backman, P. A., Rodriguez, R., and Ploper, D. L.,** Chitin as a foliar amendment to modify microbial ecology and control disease, *Phytopathology,* 81, 1152, 1991.

76. **Davis, R. F., Backman, P. A., Rodriguez-Kabana, R., and Kokalis-Burelle, N.,** Biological control of apple fruit diseases by *Chaetomium globosum* formulation containing cellulose, *Biol. Control,* 2, 118, 1992.

77. **Tronsmo, A.,** Effect of fungicides and insecticides on growth of *Botrytis cinera, Trichoderma viride* and *T. harzianum, Norw. J. Agric. Sci.,* 3, 151, 1989.

78. **Lindemann, J.,** Genetic manipulation of microorganisms for biological control, in *Biological Control on the Phylloplane,* Windels, C. E. and Lindow, S. E., Eds., APS Press, St. Paul, MN, 1985, 116.

79. **Baker, R.,** Molecular biology in control of fungal pathogens, in *Handbook of Applied Mycology. Vol. 1, Soil and Plants,* Arora, D. K., Rai, B., Mukerji, K. G., and Knudsen, G. R., Eds., Marcel Dekker, New York, 1991, 259.

80. **Doyle, M. P. and Marth, E. H.,** Food safety issues in biotechnology, in *Agricultural Biotechnology. Issues and Choices,* Baumgardt, B. R. and Martin, M. A., Eds., Purdue University Agricultural Experimental Station, West Lafayette, IN, 1991, 55.

81. **Rogoff, M. H.,** Regulatory safety data requirements for registration of microbial pesticides, in *Microbial and Viral Pesticides,* Kurstak, E., Ed., Marcel Dekker, New York, 1982, 645.

82. **Brackett, R. E.,** Shelf stability and safety of fresh produce as influenced by sanitation and disinfection, *J. Food Prot.,* 55, 808, 1992.

83. **Madden, J. M.,** Microbial pathogens in fresh produce — the regulatory perspective, *J. Food Prot.,* 55, 821, 1992.

84. **von Graevenitz, A.,** The role of opportunistic bacteria in human disease, *Annu. Rev. Microbiol.,* 31, 447, 1977.

85. **Schroth, M. N., Cho, J. J., Green, S. K., and Kominos, S. D.,** Epidemiology of *Pseudomonas aeruginosa* in agricultural areas, in *Pseudomonas aeruginosa: Ecological Aspects and Patient Colonization,* Young, V. M., Ed., Raven Press, New York, 1977, 1.

40

86. **Stoddart, J. C.**, Hospital-acquired infections, in *Care of the Critically Ill Patient*, Tinker, J. and Rapin, M., Eds., Springer-Verlag, New York, 1983, 873.

87. **Kominos, S. D., Copeland, C. E., and Delenko, C. A.**, *Pseudomonas aeruginosa* from vegetables, salads, and other foods served to patients with burns, in *Pseudomonas aeruginosa: Ecological Aspects and Patient Colonization*, Young, V. M., Ed., Raven Press, New York, 1977, 59.

88. **Neu, H. C.**, Ecology, clinical significance, and antimicrobial susceptibility of *Pseudomonas aeruginosa*, in *Non-Fermentative Gram-Negative Rods*, Gilardi, G. L., Ed., Marcel Dekker, New York, 1985, 11.

89. **Chun, D. and McDonald, R. E.**, Seasonal trends in the population dynamics of fungi, yeasts, and bacteria on fruit surface of grapefruit in Florida, *Proc. Fla. State Hortic. Soc.*, 100, 23, 1987.

90. **Lievens, K. H., Van Rijsbergen, R., Leyns, F. R., Lambert, B. J., Tenning, P., Swings, J., and Joos, H. J. P.**, Dominant rhizosphere bacteria as a source for antifungal agents, *Pestic. Sci.*, 27, 141, 1989.

91. **Farooqi, W. A., Shaukat, G. A., Malik, M. A., and Ahmad, M. S.**, Studies on the inhibition of *Alternaria citri* in stored citrus fruits, *Proc. Fla. State Hortic. Soc.*, 94, 279, 1981.

92. **Janisiewicz, W. J., Yourman, L., Roitman, J., and Mahoney, N.**, Postharvest control of blue mold and gray mold of apples and pears with pyrrolnitrin, a metabolite of *Pseudomonas cepacia, Plant Dis.*, 75, 490, 1991.

93. **Appel, D. J., Gees, R., and Coffey, M. D.**, Biological control of the postharvest pathogen *Penicillium digitatum* on Eureka lemons, *Phytopathology*, 78, 1593, 1988.

94. **Nishida, M., Matsubara, T., and Watanabe, N.**, Pyrrolnitrin, a new antifungal antibiotic, *J. Antibiot.*, 18, 211, 1965.

95. **di Menna, M. E., Mortimer, P. H., and White, E. P.**, The genus *Myrothecium*, in *Mycotoxic Fungi, Mycotoxins, and Mycotoxicoses*, Vol. 1, Wyllie, T. D. and Morehouse, L. G., Eds., Marcel Dekker, New York, 1977, 107.

96. **Rammelsburg, M. and Radler, F.**, Antibacterial polypeptides of *Lactobacillus* species, *J. Appl. Bacteriol.*, 69, 177, 1990.

97. **Batish, V. K., Grover, S., and Lal, R.**, Screening lactic acid starter cultures for antifungal activity, *Cultured Dairy Prod. J.*, 24, 21, 1989.

98. **Gutterson, N.**, Microbial fungicides: recent approaches to elucidating mechanisms, *Crit. Rev. Biotechnol.*, 10, 69, 1990.

99. **Phillips, D. J. and Harvey, J. M.**, Selective medium for detection of inoculum of *Monilinia spp.* on stone fruits, *Phytopathology*, 65, 1233, 1975.

100. **Bryde, R. J. W. and Willetts, H. J.**, *The Brown Rot Fungi of Fruit: Their Biology and Control*, Pergamon Press, New York, 1977.

101. **Jenkins, P. T. and Reinganum, C.**, The occurrence of a quiescent infection of stone fruits caused by *Sclerotinia fructicola* (Wint.) Rehm., *Aust. J. Agric. Res.*, 16, 131, 1965.

102. **Ogawa, J. M., Manji, B. T., and Sonoda, R. M.**, Management of the brown rot disease on stone fruits and almonds in California, in *Proc. Brown Rot of Stone Fruit Workshop*, Special Report No. 55, New York Agricultural Experimental Station, 1985, 8.

103. **Ogawa, J. M.**, Pre- and postharvest decay of fresh market peaches, plums and nectarines, *Calif. Fruit Grower*, 64, 16, 1987.

104. **Good, H. M. and Zathureczky, P. G. M.**, Effects of drying on the viability of germinated spores of *Botrytis cinerea, Cercospora musae*, and *Monilinia fructicola*, *Phytopathology*, 57, 719, 1967.

105. **Ogawa, J. M., English, H., Moller, W. J., Manji, B. T., Rough, D., and Koike, S. T.**, *Brown Rot of Stone Fruits*, University of California Cooperative Extension Leaflet 2206, 1980.

106. **Jerome, S. M. R.,** Brown rot of stone fruits. Latent contamination in relation to spread of the disease, *J. Aust. Inst. Agric. Sci.,* 24, 132, 1958.

107. **Philips, D. J.,** Effect of temperature on *Monilinia fructicola* conidia produced on fresh stone fruits, *Plant Dis.,* 68, 610, 1984.

108. **Tate, K. G. and Ogawa, J. M.,** Nitidulid beetles as vectors of *Monilinia fructicola* in California stone fruits, *Phytopathology,* 65, 977, 1975.

109. **Sommer, N. F.,** Suppressing postharvest disease with handling practices and controlled environments, in *Peaches, Plums, and Nectarines, Growing and Handling for Fresh Market,* LaRue, J. H. and Johnson, R. S., Eds., University of California Publication 3331, 1989, 179.

110. **De Cal, A., Sagasta, E. M., and Melgarejo, P.,** Biological control of peach twig blight (*Monilinia laxa*) with *Penicillium frequentans, Plant Pathol.,* 39, 612, 1990.

Chapter 4

The Postharvest Environment and Biological Control

Robert A. Spotts and Peter G. Sanderson

CONTENTS

I. INTRODUCTION

Many challenges must be met before biocontrol can be used successfully on a commercial basis for control of decay of fruits and vegetables. In the following discussion, examples of the types of the physical environments that may be encountered in diverse industries are presented. This is followed by a section describing the postharvest environment of pome fruit in the U.S. Pacific Northwest with specific reference to biological control. The technologies and techniques that are presently employed for produce handling must be taken into consideration when developing effective biological control strategies. It is hoped that the principles presented can be applied to a wide range of crops and production areas.

The tremendous diversity of crop/pathogen/environment combinations present in the postharvest arena presents a challenge to the commercialization of biological control. Biocontrol agents must be resistant to chemicals used for control of fungal and bacterial diseases, as well as physiological disorders. Biocontrol agents must be compatible with commercial handling systems, including dump tanks and flumes, drenches, line spray applicators, and heat tunnels. In addition, biocontrol must be effective in a wide range of temperatures and storage atmospheres. With proper planning and innovative research, these challenges can be met.

One of the keys to the successful commercialization of biological control of postharvest diseases lies in understanding and controlling the postharvest environment. In general, discussions concerning the effect of environment on the development of disease refer only to the physical environment, centering on such things as the influences of temperature, free moisture, and relative humidity on disease control. However, this discussion largely concerns the biotic environment, which includes both pathogenic and antagonistic microorganisms. The postharvest environment may be one in which the best chance to develop successful biological controls exists because many aspects of it can be controlled.

For a disease relationship to develop, a virulent pathogen, susceptible host, and conducive environment must occur concurrently in time and space. A number of postharvest pathogens such as those that cause blue and gray mold are ubiquitous. With the assumption that both a virulent pathogen or pathogens and susceptible hosts are present in the postharvest environment, several disease management strategies can be developed. The usual approach has been to treat the produce with fungistatic or fungicidal chemicals. Other strategies that can be employed include manipulating the physical environment to make it nonconducive to disease development and introducing biological agents that compete against or hyperparasitize the pathogen. In addition to affecting the host/pathogen relationship directly, modification of the physical environment may be implemented to enhance the activity of the antagonist against a targeted pathogen.

II. PHYSICAL ENVIRONMENT

Modified environment techniques have significantly improved the control of decay in fruits and vegetables. Commercial producers and handlers modify temperature, relative humidity, and atmospheric composition during prestorage, storage, and transit to control decay. For optimum decay control, two or more environmental factors are often modified simultaneously. Although temperature is probably the most commonly controlled environmental factor for preventing postharvest decay, these other factors may interact favorably to control decay. In addition, research has been conducted to investigate the potential for use of low atmospheric pressure (hypobaric atmosphere), and irradiation (gamma and ultra violet) for decay control. With technological advances in instrumentation and storage facilities, refinements of existing practices and development of new methods for postharvest disease control are continuously being made. Excellent reviews of this area have been published.[1,2]

A. TEMPERATURE

Many factors affect the rate and ability to cool produce and include type and size of storage container, stacking patterns, air flow, relative humidity, actual temperature inside bins and fruit, and respiration rates and heat evolution.[1] Delays between harvest and cold storage, as well as temperature pull-down rate, may affect decay control. As time between harvest and storage is lengthened, *Penicillium* decay of apple increases.[3] Decay of grapes, however, does not differ between short and long precooling times followed by storage at -0.5°C.[4] It is important to keep storage temperature variation at a minimum because seemingly small differences in temperature can drastically affect storage life. For example, the storage life of apples and pears is more greatly increased by storage at -1°C than at 0°C (25 and 40%, respectively).[1]

Generally, storage of fruits and vegetables at temperatures just above the injury threshold minimizes decay. Optimal storage temperatures can range from -1.7°C for pears to 21.1°C for tomatoes.[1] Chilling and freezing stress often alter metabolism, resulting in weakened or injured tissue, and predispose fruits and vegetables to attack by pathogenic microorganisms. For example, eggplants are injured at about 5°C and become highly susceptible to *Alternaria* rot.[5] Sweet peppers are injured below 7°C and become susceptible to *Alternaria* and *Botrytis* decay.[6-8] Tomatoes stored below 10°C are susceptible to *Alternaria* rot, and susceptibility increases with time of cold exposure.[9] Below 10°C, sweet potatoes are predisposed to *Alternaria, Botrytis, Mucor,* and *Penicillium* decay.[10] Carrots stored *in situ*, when not mulched, are subject to freezing injury and susceptible to rot.[11]

Many decay fungi can grow below 5°C and cannot be controlled solely with low temperature storage. *Coprinus psychromorbidus* Redhead and Traquair, a low temperature basidiomycete causing decay of Anjou pears, grows optimally at 10°C and is difficult to isolate at room temperature.[12,13] Numerous decay fungi cause problems in frozen foods, and fluctuating temperature and relative humidity during defrost cycles generally are involved.[14]

High temperature also may be used to control decay and is particularly useful for crops such as mango, papaya, pepper, and tomato that are injured by low temperature. In hot water treatments, temperature margins between decay control and crop injury are often extremely narrow. Cultivar, growing climate, and time between infection and treatment all may affect the success of hot water decay control. The higher the temperature used to treat the produce, the shorter the treatment time necessary to obtain control. Occasionally, injury does not appear immediately but may develop after several weeks in storage. Thus, while hot water can be used to control *Monilinia* and *Rhizopus* decays of fresh market peaches, it cannot be used on fruit that is to be stored after treatment.[15] Decay control may be effective at lower water temperatures if biological or chemical fungicides are to be added to the hot water.[16,17]

Although hot water generally is more effective, hot air has been used to control decay in crops that are injured in hot water. For example, *Botrytis* and *Rhizopus* decays of strawberry[18,19] and raspberry[20] are controlled by exposure to humid, 44°C air for 30 to 60 min. During a 3- to 10-d curing period, formation of wound periderm in sweet potato (*Ipomoea batatas*) increases as temperature is raised from 12 to 32°C, provided relative humidity is above 90%.[21] The wound periderm serves to protect the tuberous root from desiccation and fungal infection. Wound healing of potato tubers (*Solanum tuberosum*) is promoted by curing for about 2 weeks under conditions of high humidity and good air circulation at 13°C.[22] Various decays of apples[23] and pumpkins[24] have been controlled by prestorage hot air treatment at 38° and 26°C, respectively, for approximately 1 week.

B. RELATIVE HUMIDITY

Both low and high relative humidity (RH) have been related to postharvest decay control. Ambient cold room RH often differs from that within fruit or vegetable containers or wound microsites. Perforated polyethylene bags for fruit and vegetable storage create RH about 5 to 10% above that in the storage room. Although shrivel and weight loss are reduced with polyethylene bags, decay may be increased.

Crops such as leafy vegetables store better at high (98 to 100%) RH and subsequently decay less because fewer dead and chlorotic tissues, which may provide nutrients for decay organisms, accumulate.[25-27] In addition, production of fungal enzymes may be inhibited in high RH, and deactivation of enzymes may be more rapid.[26]

In contrast, crops such as apples and pears with well-developed cuticles and epidermal layers tolerate relatively lower RH levels, which helps to prevent storage decay. Often fungal spore germination is inhibited at low RH, and small differences in RH can have significant effects. At -1.1°C, for example, *Mucor piriformis* Fischer, *Botrytis cinerea* Pers.: Fr., *Pezicula malicorticis* (Jacks.) Nannf., *Penicillium expansum* Link, and *Phialophora malorum* (Kidd and Beaum.) McCulloch germinate at 100% RH in the absence of free water, but none of these fungi germinate at 97% RH.[28] However, when wounded Anjou pears were inoculated with conidia of these five fungi and placed at 97, 99, or 100% RH at -1.1°C, *B. cinerea*, *Penicillium expansum*, and *Pezicula malicorticis* caused decay at all RH levels; however, *M. piriformis* and *Phialophora malorum* caused decay only at 100% RH.[28]

C. CONTROLLED ATMOSPHERE

Alterations in concentration of oxygen (O_2), carbon dioxide (CO_2), nitrogen (N_2), carbon monoxide (CO), and ethylene can significantly affect hosts, pathogens, and biocontrol agents. Controlled atmospheres (CA) are advantageous for decay control of some fruits and vegetables, but produce negative effects on others. Gas concentrations and ratios must be carefully controlled to avoid undesirable effects.

Changes in O_2 and CO_2 concentrations are the most common type of CA, with reduction of O_2 usually accompanied by a corresponding increase in N_2. The following examples of decay control with CA illustrate the value of this technique. Cox's Orange Pippin apples store well in 3% O_2 + 5% CO_2, *Gleosporium album* Osterw. pectolytic enzyme production decreases, and decay is controlled.[29] Both anthracnose and chilling injury of avocados are reduced in 2% O_2 + 10% CO_2.[30] *Botrytis porri* infection of leeks is reduced in 1% O_2 + 10% CO_2.[31] Stem-end decay of Anjou pears is reduced when fruit are stored in 1% O_2 + 99% N_2.[32] Tomatoes stored in 3% O_2 + 3% CO_2 show less anthracnose, *Penicillium, Botrytis,* and bacterial soft rot than those in air storage; however, *Fusarium* rot increases if RH is 95 to 100%.[33] Control of decay in chili peppers,[34] muskmelons,[35] and prunes[36] has been reported in 10 to 30% CO_2. Generally, high CO_2 levels appear to be more effective for control of fungal than of bacterial diseases. The CO_2 is less effective when the nutrient supply for pathogens is increased by wounding.[37]

Interactions between CA and temperature are not uncommon. In citrus, incidence of decay declines as O_2 is lowered from 21 to 10%, but only at 1°C.[38] Soft rot of tomato is reduced in 3% O_2 + 5% CO_2 at 13°C, but not at 7 or 18°C.[39]

Often ethylene accumulates in CA-stored fruits and vegetables and affects decay. Removal of ethylene resulted in reduced anthracnose of avocados[40] and decreased *Penicillium* and *Diaporthe* decay of lemons[41] stored in CA. In contrast, *Gleosporium album* decay of CA-stored apples was reduced in high ethylene; however, *in vitro* growth of the pathogen was not correspondingly inhibited.[42]

Several mechanisms are involved in decay control with CA. Control of pear stem-end decay in 1% O_2 apparently is primarily related to improved stem condition.[32] In addition to delaying host senescence, CA may also directly affect decay organisms. Generally, growth of *Botrytis, Monilinia,* and *Penicillium* is reduced when the concentration of O_2 is less than 2%. In 1% O_2, *Botrytis* does not sporulate *in vitro*.[43] *Rhizopus* produce no mature sporangia in 0.5% O_2,[43] and *Rhizopus* decay of strawberries decreases linearly as O_2 is lowered from 21 to 0%.[44] *Rhizopus*, however, can grow in 100% N_2.[45] *In vitro* growth and sporulation of *Botrytis, Penicillium,* and *Rhizopus* decrease as CO_2 is raised from ambient to 10.5%;[46] however, conidia of *Penicillium* can still germinate in 60% CO_2.[37]

Effects of CA on pathogens *in vitro* do not always correspond to effects of CA on decay by those pathogens. In 7.5 to 15% O_2, *Gleosporium album* growth *in vitro* is depressed, but decay of apples by it is increased.[47] Growth of the same pathogen is stimulated at 5 to 10% CO_2 *in vitro*, but decay of apples is reduced in 5% CO_2 atmosphere.

III. POME FRUIT INDUSTRY IN THE PACIFIC NORTHWEST

A. CROP DIVERSITY AND DECAY RESISTANCE

In the Pacific Northwest, many of the large fruit packinghouses handle several crops such as apples, pears, and sweet cherries. In the U.S., Red Delicious is the leading apple cultivar with a yield of over 97 million bushels (1 bushel $\approx$ 20 kg), but production of nine other cultivars (e.g., Golden Delicious, McIntosh, Newtown, etc.) exceeds 4 million bushels (USDA and IAI estimate, 1989). At present, there is a great deal of interest in a number of relatively new varieties such as Gala, Braeburn, and Fuji. While some of the earliest plantings of these varieties are just beginning to come into full production, new plantings are being established. Similarly, the pear cultivars Bartlett, Anjou, Bosc, and Comice — as well as several others — are grown in large quantities. As with new apple cultivars, many new plantings of red-colored varieties of Bartlett and Anjou are being established or are just coming into production. In addition, several cultivars of sweet cherry are grown in significant quantities. Cultivars vary in resistance to decay, and each may require a different biocontrol strategy. For example, the pear cultivars Bosc and Comice are highly susceptible to side rot caused by *Phialophora malorum*, but Anjou is relatively resistant.[48] Preliminary tests in our laboratory have shown that there is a considerable difference in the rate of decay by *Botrytis cinerea, Penicillium expansum,* and *Mucor piriformis* among different apple cultivars.

B. PATHOGEN DIVERSITY

Perhaps even more striking than crop diversity is the complex group of pathogenic fungi and bacteria that cause postharvest decay of fruits and vegetables. On apples and pears, the most important postharvest diseases are blue mold, gray mold, *Mucor* rot, and bullseye rot. *Pezicula malicorticis,* the causal agent of bullseye rot, can infect fruit from petal fall to harvest.[49] *Coprinus psychromorbidus,* a low temperature basidiomycete, occasionally is a serious disease on pears and appears to infect fruit just before harvest.[12] The most common postharvest pathogens of pome fruit, *Penicillium expansum* (blue mold) and *Botrytis cinerea* (gray mold) infect fruit during and after harvest through wounds or stem ends.[50] Furthermore, several species of *Penicillium* other than *P. expansum* have been shown to be relatively common in the postharvest environment of apples and pears.[51-54] *Penicillium solitum* is of particular interest because inoculum is prevalent in packinghouses, and all of the isolates that have been recovered from fruit or fruit handling systems and tested for sensitivity to benzimidazole fungicides have been found to be insensitive.[52,53,55]

Dose/response relationships have been determined for a number of postharvest pome fruit decay organisms, which tend to show a monomolecular or simple interest curve. On newly harvested McIntosh apples, disease incidence due to infection by *P. expansum* showed an asymptote at about 2.0 10^3 conidia per milliliter, resulting in about 50% infection after which 90% infection was reached with 1.5×10^4 conidia per milliliter.[56] In the same study, *Botrytis* showed a more linear relationship with a much lower response. About 30% infection resulted from inoculations using 2.0×10^3 conidia per milliliter. Disease incidence in Anjou pears that were inoculated with conidia of *P. expansum, B. cinerea,* and *M. piriformis* also showed a dose/response with curves that had an asymptote at about 1.5×10^3 conidia per milliliter.[57] In wounded fruit at that inoculum concentration, *P. expansum* caused the highest incidence of decay followed by *B. cinerea* and *M. piriformis* (approximately 75, 55, and 35%, respectively).[57] Stem-end decay, however, was greatest on fruit inoculated with *M. piriformis* followed by *B. cinerea* and *P. expansum* (approximately 40, 23, and 5%, respectively).[57] These numbers of pathogen spores are commonly found in packinghouse dump tanks.[52,56,58-60]

Host susceptibility to infection may increase over the storage season. Blue mold disease incidence on McIntosh apples resulting from infection by 100 conidia per milliliter of *P. expansum* increased from 30% to about 100% from September to April, respectively; and disease incidence on McIntosh apples increased from about 10 to 90% after infection by 250 conidia per milliliter of *B. cinerea.*[56]

It is unlikely that any single biocontrol agent will control such a diverse group of pathogens. For example, in tests conducted in our laboratory with a bacterium that showed promise as a potential biological control agent, good control of blue mold and moderate control of gray mold were achieved, but the bacterium had no effect on *Mucor* rot. In further tests with that and another species of bacteria and the yeast, *Cryptococcus laurentii,* we found that control of *P. expansum* broke down with pathogen inoculum concentrations greater than 2.0×10^3 conidia per milliliter. In addition, it is uncertain that postharvest application of biocontrol agents will eradicate latent infections caused by *P. malicorticis* and *C. psychromorbidus*.

C. COMPATIBILITY WITH PREHARVEST AND POSTHARVEST CHEMICALS

Several fungicides with different modes of action may be applied within the last 7 d before harvest for control of late season/storage scab and postharvest decay. Thus, biocontrol agents applied postharvest would contact fungicide residues on fruit surfaces. In addition, several fungicides and antiscald chemicals (diphenylamine for apples, ethoxyquin for pears) are applied to fruit in postharvest drenches or line sprays.[61] Although the use of effective biocontrols would reduce or eliminate the need for some chemical applications, biocontrol agents must be selected with resistance to the chemicals commonly used in fruit and vegetable production.

Application of a preharvest spray for control of postharvest decay currently is recommended.[61] On apples, ziram and captan may be applied up to 0 and 4 days, respectively, before harvest.[62,63] Only ziram is registered for postharvest decay control on pears. In addition, triadimefon may be applied up to 7 d before harvest for mildew control and dodine up to 0 days before harvest for late season/storage scab.

1. Drenches and Line Sprays

Apples and pears are often treated with fungicides and antiscald chemicals in postharvest drenches (Figure 1) or line sprays (Figure 2). Diphenylamine and ethoxyquin (antiscald treatments for apples and pears, respectively) are usually mixed with either thiabendazole or captan, the only products currently registered for postharvest use. Usually only apples and pears that are to be placed directly in CA storage without presizing or packing into shipping boxes are drenched. In a recent survey it was found that about 32% of packinghouses routinely drench Anjou pears before storage.[64]

Line sprays are used to deliver fungicides, antiscald agents, waxes, and other materials to the fruit after it has been removed from the field bin. The different materials may or may not be mixed prior to application. Several different types of applicators are in use. Materials may be sprayed directly onto the fruit primarily with cone-type nozzles using either hydraulic or air pressure for delivery. Some packinghouses simply drip the materials onto either natural or synthetic brushes that in turn coat the fruit with the chemical. Recent interest has been shown in the use of controlled droplet applicators that use a propeller to shear droplets of the material to be applied into a fine mist. Devices that place an electrostatic charge on the droplets are being tested in conjunction with the controlled droplet applicators and appear to be effective in increasing the amount of residue on the fruit.

2. Dump Tanks

To minimize injury to fruit when it is removed from field bins to be sorted, sized, or packed, field bins are submerged in dump tanks containing water or a salt solution; and the fruit is floated out of the bins (Figure 3). The specific gravity of apples is such that they will float in water alone; however, but pears are more dense, and flotation salts must be added to the water in the dump tank for the pears to float sufficiently. Salts are added to the dump tank water to bring the specific gravity up to about 1.02 to 1.05. Several different salts may be used, and they generally are combined with a disinfestant to kill decay pathogen propagules in the flotation solution. Sodium lignin sulfonate, a by-product of the paper pulping industry, is the most commonly used flotation salt (ca. 50% of packinghouses), followed by sodium silicate (30%) and sodium sulfate (20%).[64] Other

Figure 1 Drench solution containing fungicides and antiscald agents being applied to fruit in wooden bins at a packinghouse in the U.S. Pacific Northwest. Large tank (top) at right contains the drench solution, which is under constant agitation. The solution is caught in the drive-through trough and recirculated. Bins are slotted on the bottom and material can pass from upper to lower bins. In addition to flooding the bins with drench solution from the top, nozzles direct spray to openings between bins at the sides (bottom), ensuring thorough coverage.

Figure 2 Line spray apparatus in a presize line at a packinghouse in the U.S. Pacific Northwest. Fungicides are applied to fruit as it passes through the tent (top) after fruit is sorted to remove culls. Waxes and antiscald agents also may be applied at this point. Inside the tent (bottom), a controlled droplet applicator is shown applying thiabendazole to Bosc pears. As droplets of the fungicide suspension fall onto the blades of the small propeller, seen in the center of the photograph, they are sheared into a fine mist.

Figure 3 Field bins containing Bosc pears being immersed into a dump tank in the U.S. Pacific Northwest. Fruit is floated out of bins to avoid damage from bruises and punctures. Debris carried into the tank on fruit and especially on bins may harbor propagules of decay organisms. Therefore, disinfestants such as chlorine or sodium *o*-phenylphenate typically are added to the water in the tank. In pear lines, such as this one, flotation salts are added to increase the specific gravity of the water to aid in removal of the fruit.

salts that may be used for flotation are sodium carbonate and calcium lignin sulfonate.[61] The most common disinfestants used are sodium *o*-phenylphenate (SOPP) and chlorine, the latter as sodium hypochlorite or calcium hypochlorite. It is recommended that chlorine levels be maintained at 100 μl/ml and SOPP at 0.3 to 0.5%[61] in commercial packinghouses. Chlorine is not compatible with the lignin sulfonates because it reacts with the lignin.

The various salt solutions differ in pH, which affects the efficacy of the disinfestant used. The efficacy of chlorine is increased as the pH is reduced from 11–12 to pH 8, but at lower pH the chlorine becomes unstable.[65] A pH range of 8.0 to 8.5 has been recommended as a balance between high activity and stability.[66,67] Sodium carbonate and sodium silicate solutions have a pH of about 11.2 with either chlorine or SOPP.[68] The pH of sodium sulfate solutions with chlorine is about 7.8 and with SOPP rises to about 10.8.[68] The pH of the lignin sulfonate solutions is about 6.7 and 6.4 for the sodium and calcium salts, respectively; and increases to about 8.7 and 7.8, with the addition of SOPP, respectively.[68]

Other dump tank disinfestants are being investigated in an attempt to find alternatives to chlorine and SOPP. Ozone is being used as a disinfestant by a few packinghouses in the Pacific Northwest. LD_{95} values for spores of *B. cinerea, M. piriformis*, and *P. expansum* were 0.99, 0.69, and 0.39 μg of ozone per milliliter of water, respectively.[69] Research also has been conducted with chlorine dioxide, but worker safety issues have subdued enthusiasm toward this material for indoor use.

D. FRUIT HANDLING SYSTEMS

At harvest, fruit are placed directly into wooden field bins that hold about 450 kg of fruit. These bins may be drenched with fungicides and an antiscald chemical and stored in air or CA storage. However, in a presize system, fruit are floated from bins, sorted, treated with a fungicide in water or wax, sized, and then placed back into bins for storage. Alternatively, fruit may be packed and stored in boxes, each containing about 20 kg of fruit rather than in bins.

Recycled drench solutions accumulate soil and debris containing fungal spores. These solutions usually are changed every 1 to 7 d, with drencher volumes commonly exceeding 4000 l. Application of biocontrol agents in this system would require considerable quantities of material.

Immersion dump tank and flume systems are used to remove fruit from bins and move it through the packinghouse. Solution volume often exceeds 12,000 l. A flotation salt such as sodium sulfate or sodium lignin sulfonate is necessary to float pears but not apples. Temperature of these solutions usually is under 10°C but may be heated to 25°C to clean apples for waxing. Some packinghouses use a heat sterilization method for dump tank water. Water is heated to 55°C for 25 min, then allowed to cool before fruit is added.[59] The number of decay spores also is reduced by chlorination or use of sodium o-phenylphenate.[58] Other methods that are being used to a limited extent in fruit packinghouses in the Pacific Northwest to reduce decay spores include ozonation and filtration. The application of biocontrol agents in these types of systems has not been tested.

Fungicides often are applied to fruit in a line spray in water or wax. These systems are nonrecirculating but require only small volumes of solution. A wax solution of 1 l will cover over 1300 kg of fruit. Waxed fruit often is dried in a heat tunnel at 50 to 60°C. Biocontrol agents could be incorporated easily into a line spray system.

E. FRUIT STORAGE SYSTEMS

Apples are stored at -1 to 4°C and pears at -1.1 to -0.5°C.[1] In the Pacific Northwest, fruit is stored up to 9 months. Fruit are shipped to market at transit temperatures of 0 to 2°C. Winter pears require ripening at 20°C for 5 to 10 d. Thus, biocontrol agents must function in a wide temperature range to give control of decay from storage to the consumer.

Controlled atmosphere (CA) is required for long-term cold storage of apples and pears. Anjou pears can be stored 7 months in air if they are pretreated with 12% CO_2, 8 months in 2% O_2 + 1% CO_2, or 9 months in 1% O_2 + 0.1% CO_2.[70] Atmosphere of less than 2% O_2 often has adverse effects on microorganisms. These effects can be used advantageously to reduce decay.[32] However, biocontrol agents must tolerate a low O_2-high CO_2 environment if they are to provide decay control in commercial CA storage.

IV. CONCLUSIONS

Many biological and physical factors must be considered when selecting or developing biocontrol agents for commercial use. Packinghouse personnel are accustomed to broad spectrum fungicides that function in a wide range of postharvest handling and storage systems. Mistakes are likely to occur in making the switch to living biocontrol agents that are more sensitive to environmental conditions than are fungicides or bactericides. If biocontrol agents are too narrow in their pathogen-crop-environment action spectrum, companies may hesitate to make the financial commitments necessary for registration or implementation. These challenges must be met if biocontrol for reduction of postharvest decay of fruits and vegetables is to succeed on a commercial scale.

REFERENCES

1. **Hardenburg, R. E., Watada, A. E., and Wang, C. Y.**, *The Commercial Storage of Fruits, Vegetables, and Florist and Nursery Crops*, Agricultural Handbook 66, Agricultural Research Service, U.S. Department Of Agriculture, Washington, D.C., 1986.
2. **Salunkhe, D. K. and Desai, B. B.**, *Post Harvest Technology of Fruits*, Vol. 1, CRC Press, Boca Raton, FL, 1984.
3. **Brooks, C. and Cooley, J. S.**, Temperature relations of apple-rot fungi, *J. Agric. Res.*, 8, 139, 1917.
4. **Combrink, J. C., Laszlo, J. C., Truter, A. B., and Bosch, J. J. C.**, Can the precooling time for table grapes be shortened?, *Deciduous Fruit Grower*, 28, 166, 1978.
5. **McColloch, L. P.**, *Chilling Injury of Eggplant Fruits*, Marketing Research Report No. 749, Agricultural Marketing Service, U.S. Department Of Agriculture, Washington, D.C., 1966.
6. **Abdel-Maksoud, M., Abou Aziz, A. B., Abdel-Kadel, A. S., and Abdel Samei, K. A.**, Effect of temperature on quality and decay percentage of "California Wonder" pepper fruits grown in "Gharbia" and "Menofia" governorate during storage, *Egypt. J. Hortic.*, 2, 157, 1975.
7. **McColloch, L. P.**, *Chilling Injury and Alternaria Rot of Bell Peppers*, Marketing Research Report No. 536, Agricultural Marketing Service, U.S. Department of Agriculture, Washington, D.C., 1962.
8. **McColloch, L. P. and Wright, W. R.**, *Botrytis Rot of Bell Peppers*, Marketing Research Report No. 754, Agricultural Marketing Service, U.S. Department of Agriculture, Washington, D.C., 1966.
9. **McColloch, L. P. and Worthington, J. T.**, Low temperature as a factor in the susceptibility of mature-green tomatoes to alternaria rot, *Phytopathology*, 42, 425, 1952.
10. **Lauritzen, J. I.**, Some effects of chilling temperatures on sweet potatoes, *J. Agric. Res.*, 42, 617, 1931.
11. **Tucker, W. G. and Cox, C. J.**, Studies on the overwinter losses of carrots stored *in situ*, *J. Hortic. Sci.*, 53, 291, 1978.
12. **Spotts, R. A., Traquair, J. A., and Peters, B. B.**, d'Anjou pear decay caused by a low temperature basidiomycete, *Plant Dis.*, 65, 151, 1981.
13. **Spotts, R. A.**, Coprinus rot, in *Compendium of Apple Diseases*, Jones, A. L. and Aldwinckle, H. S., Eds., APS Press, St. Paul, MN, 1990, 59.
14. **Gunderson, M. F.**, Mold problem in frozen foods, in *Proc. Low Temperature Microbiol. Symp.*, Campbell Soup Co., Camden, NJ, 1961, 299.
15. **Smith, H. W.**, Nonchemical control of postharvest deterioration of fresh produce, in *Proc. 3rd Int. Biodegradation Symp.*, Sharpley, J. M. and Kaplan, A. M., Eds., Applied Science Publishers, London, U.K., 1976, 577.
16. **Wells, J. M.**, Postharvest Hot-water and Fungicide Treatments for Reduction of Decay of California Peaches, Plums, and Nectarines, Marketing Research Report No. 980, Agricultural Marketing Service, U.S. Department of Agriculture, Washington, D.C., 1971.
17. **Wells, J. M. and Harvey, J. M.**, Combination heat and 2,6-dichloro-4-nitroaniline treatments for control of *Rhizopus* and brown rot of peaches, plums, and nectarines, *Phytopathology*, 60, 116, 1970.
18. **Couey, H. M. and Follstad, M. N.**, Heat pasteurization for control of postharvest decay in fresh strawberries, *Phytopathology*, 56, 1345, 1966.
19. **Smith, W. L., Jr. and Worthington, J. T., III**, Reduction of postharvest decay of strawberries with chemical and heat treatments, *Plant. Dis. Rep.*, 49, 619, 1965.

20. **Worthington, J. T., III and Smith, W. L., Jr.,** Postharvest decay control of red raspberries, *Plant Dis. Rep.,* 49, 783, 1965.
21. **Lauritzen, J. I.,** Factors affecting infection and decay of sweet potatoes, *J. Agric. Res.,* 50, 285, 1935.
22. Anon., *Potato Information Sheet,* Sheet No. G-1, University of Maine Cooperative Extension Service, Orono, ME, 1969.
23. **Porritt, S. W. and Lidster, P. D.,** The effect of prestorage heating on ripening and senescence of apples during cold storage, *J. Am. Soc. Hortic. Sci.,* 103, 584, 1978.
24. **Gorini, F. L., and Testoni, A.,** Conservazione a lungo termine delle zucche, *Ann. Isti. Sper. Valorizzazione Tecnol. Prod. Agric.,* 9, 131, 1978.
25. **Pendergrass, A. and Isenberg, F. M. R.,** The effect of relative humidity on the quality of stored cabbage, *HortScience,* 9, 226, 1974.
26. **van den Berg, L. and Lentz, C. P.,** High humidity storage of some vegetables, *Can. Inst. Food Sci. Technol. J.,* 7, 260, 1974.
27. **Yoder, O. C.,** Development of methods for long-term cabbage storage, *Acta Hortic.,* 62, 301, 1977.
28. **Spotts, R. A. and Peters, B. B.,** The effect of relative humidity on spore germination of pear decay fungi and d'Anjou pear decay, *Acta Hortic.,* 124, 75, 1981.
29. **Edney, K. L.,** The effect of the composition of the storage atmosphere on the development of rotting of Cox's Orange Pippin apples and the production of pextolytic enzymes by *Gloeosporium* spp., *Ann. Appl. Biol.,* 54, 327, 1964.
30. **Spalding, D. H. and Reeder, W. F.,** Low-oxygen high-carbon dioxide controlled atmosphere storage for control of anthracnose and chilling injury of avocados, *Phytopathology,* 65, 458, 1975.
31. **Hoftun, H.,** Storage of leeks. III. Storage in controlled atmospheres, *Meld. Nor. Landbrukshoegsk.,* 57, 1, 1978.
32. **Chen, P. M., Spotts, R. A., and Mellenthin, W. M.,** Stem-end decay and quality of low oxygen stored d'Anjou pears, *J. Am. Soc. Hortic. Sci.,* 106, 694, 1981.
33. **Lockhart, C. L., Eaves, C. A., and Chipman, E. W.,** Suppression of rots on four varieties of mature green tomatoes in controlled atmosphere storage, *Can. J. Plant Sci.,* 49, 265, 1969.
34. **Kader, A. S., Klaustermeyer, J. A., Morris, L. L., and Kasmire, R. F.,** Extending storage-life of mature-green chili peppers by modified atmospheres, *HortScience,* 10 (Abstr.), 43, 1975.
35. **Stewart, J. K.,** Decay of muskmelons stored in controlled atmospheres. *Sci. Hortic.,* 11, 69, 1979.
36. **Brooks, C., Miller, E. V., Bratley, C. O., Cooley, J. S., Mook, P. V., and Johnson, H. B.,** *Effect of Solid Gaseous Carbon Dioxide Upon Transit Diseases of Certain Fruits and Vegetables,* Technical Bulletin 318, U.S. Department of Agriculture, Washington, D.C., 1932.
37. **Smith, H. W.,** The use of carbon dioxide in the transport and storage of fruits and vegetables, *Adv. Food Res.,* 12, 95, 1963.
38. **Smoot, J. J.,** Decay of Florida fruit stored in controlled atmospheres and in air, in *Proc. 1st Int. Citrus Symp.,* 3, 1285, 1969.
39. **Parsons, C. S. and Spalding, D. H.,** Influence of a controlled atmosphere, temperature, and ripeness on bacterial soft rot of tomatoes, *J. Am. Soc. Hortic. Sci.,* 97, 297, 1972.
40. **Hatton, T. T., Jr. and Reeder, W. F.,** Quality of 'Lula' avocados stored in controlled atmospheres with or without ethylene, *J. Am. Soc. Hortic. Sci.,* 97, 339, 1972.
41. **Wild, B. L., McGlasson, W. B., and Lee, T. H.,** Effect of reduced ethylene levels in storage atmospheres on lemon keeping quality, *HortScience,* 11, 114, 1976.

42. **Lockhart, C. L., Forsyth, F. R., and Eaves, C. A.,** Effect of ethylene on development of *Gloeosporium album* in apple and on growth of the fungus in culture, *Can. J. Plant. Sci.,* 48, 557, 1968.

43. **Follstad, M. N.,** Mycelial growth rate and sporulation of *Alternaria tenuis, Botrytis cinerea, Cladosporium herbarium,* and *Rhizopus stolonifer* in low-oxygen atmospheres, *Phytopathology,* 56, 1098, 1966.

44. **Wells, J. M.,** Growth of *Rhizopus stolonifer* in low oxygen atmospheres and production of pectic and cellulolytic enzymes, *Plant. Dis. Rep.,* 58, 1598, 1968.

45. **Wells, J. M. and Uota, M.,** Germination and growth of five fungi in low-oxygen and high-carbon dioxide atmospheres, *Phytopathology,* 60, 50, 1969.

46. **Littlefield, N. A., Wankier, B. M., Salunkhe, D. K., and McGill, J. N.,** Fungistatic effects of controlled atmospheres, *Appl. Microbiol.,* 14, 579, 1966.

47. **Dewey, D. H., Herner, R. C., and Dilley, D. R.,** Effect of CA storage on storage rot pathogens, in *Proc. Natl. Controlled Atmos. Res. Conf.,* 9, 113, 1969.

48. **Sugar, D.,** The Disease Cycle of Side Rot of Pear Caused by *Phialophora malorum,* Ph.D. thesis, Oregon State University, Corvallis, OR, 1989.

49. **Spotts, R. A.,** Effect of preharvest pear fruit maturity on decay resistance, *Plant Dis.,* 69, 388, 1985.

50. **Pierson, C. F., Ceponis, M. J., and McColloch, L. P.,** *Market Diseases of Apples, Pears, and Quinces,* Agricultural Handbook 376, Agricultural Research Service, U.S. Department of Agriculture, Washington, D.C., 1971.

51. **Sanderson, P. G., Rosenberger, D. A., and Spotts, R. A.,** Occurrence, pathogenicity and competition among species of *Penicillium* on apple and pear fruit, *Phytopathology,* 81, 1175, 1991.

52. **Rosenberger, D. A., Wicklow, D. T., Korjagin, V. A., and Rondinaro, S. M.,** Pathogenicity and benzimidazole resistance in *Penicillium* species recovered from floatation tanks in apple packinghouses, *Plant Dis.,* 75, 712, 1991.

53. **Holmes, R. J.,** *An Analysis of Postharvest Losses in the Victorian Pomefruit Industry,* Research Report Series No. 115, Department of Agriculture and Rural Affairs, Knoxfield, Victoria, Australia, 1990.

54. **Sanderson, P. G. and Spotts, R. A.,** Blue mold—an expanded concept, in *Proc. 8th Annu. Washington Tree Fruit Postharvest Conf.,* Washington State Horticultural Association, Wenachee, WA, 1992, 86.

55. **Pitt, J. I., Spotts, R. A., Holmes, R. J., and Cruickshank, R. H.,** *Penicillium solitum* revived, and its role as a pathogen of pomaceous fruit, *Phytopathology,* 81, 1108, 1991.

56. **Blanpied, G. D. and Purnasiri, A.,** *Penicillium* and *Botrytis* rot of McIntosh apples handled in water, *Plant Dis. Rep.,* 52, 865, 1968.

57. **Spotts, R. A.,** Relationships between inoculum concentrations of three decay fungi and pear fruit decay, *Plant Dis.,* 70, 386, 1986.

58. **Spotts, R. A. and Cervantes, L. A.,** Populations, pathogenicity, and benomyl resistance of *Botrytis* spp., *Penicillium* spp., and *Mucor piriformis* in packinghouses, *Plant Dis.,* 70, 106, 1986.

59. **Spotts, R. A. and Cervantes, L. A.,** Effects of heat treatments on populations of four fruit decay fungi in sodium ortho phenylphenate solutions, *Plant Dis.,* 69, 574, 1985.

60. **Sholberg, P. L. and Owen, G. R.,** Populations of propagules of *Penicillium* spp. during immersion dumping of apples, *Can. Plant Dis. Surv.,* 70, 11, 1990.

61. **Spotts, R., Roberts, R., Ewart, H. W., Sugar, D., Apel, G., Willett, M., and Kupferman, E.,** Management practives to minimize postharvest decay of apples and pears, *Tree Fruit Postharvest J.,* 3, 5, 1992.

62. **Riedl, H., Spotts, R. A., Mielke, E. A., Long, L. E., Fisher, G. C., Pscheidt, J. W., and William, R.,** *1992 Pest Management Guide for Tree Fruits in the Mid-Columbia Area,* Oregon State University Extension Service, EM 8293, Corvallis, OR, 1992.

63. **Beers, E. H., Grove, G. G., Williams, K. M., Parker, R., Askram, L. R., Daniels, C., and Maxwell, T.,** 1992 Crop Protection Guide for Tree Fruits in Washington, Washington State University Cooperative Extension, Pullman, WA, 1992.

64. **Kupferman, E. M. and Burkhart, D.,** Results of a survey to determine reasons for postharvest cullage of the 1990 Anjou pear crop, in *Proc. 8th Annu. Washington Tree Fruit Postharvest Conf.*, Washington State Horticultural Association, Wenachee, WA, 1992, 81.

65. **Saur, D. B. and Burroughs, R.,** Disinfestation of seed surfaces with sodium hypochlorite, *Phytopathology*, 76, 745, 1986.

66. **Link, H. L. and Pancoast, H. M.,** Antiseptic cooling retains top quality in perishables, *Food Ind.*, 21, 737, 1949.

67. **van der Plank, J. E.,** *The Use of Hypochlorous Acid and Its Salts in Citrus Packhouses for Bleaching Sooty Blotch and as Disinfectants Against Mould*, Union South Africa Department of Agriculture Bulletin, 241, 1945.

68. **Spotts, R. A. and Cervantes, L. A.,** Evaluation of disinfestant-floation salt-surfactant combinations on decay fungi of pear in a model dump tank, *Phytopathology*, 79, 121, 1989.

69. **Spotts, R. A. and Cervantes, L. A.,** Effect of ozonated water on postharvest pathogens of pear in laboratory and packinghouse tests, *Plant Dis.*, 76, 256, 1992.

70. **Meheriuk, M., Evans, C., Talley, E., and Kupferman, E.,** Harvest maturity and storage regime for pears, *Postharvest Pomology Newsl.*, 6, 11, 1988.

Chapter 5

Preharvest Management for Postharvest Biological Control

Harold E. Moline

CONTENTS

I. INTRODUCTION

There is little doubt that biological control of postharvest diseases has the potential to be of significant benefit to the fresh fruit and vegetable industry. Consumer demand for alternatives to produce treated with synthetic chemicals presently used to control postharvest diseases is a major impetus in the search for alternative means for their control. Experimental biocontrol studies have demonstrated the potential of several nonpathogenic bacterial and fungal antagonists and natural compounds for postharvest protection.

Among the major unanswered questions is when antagonists should be applied to assure their survival and maximum benefit to the host. Produce that appears healthy when harvested may harbor latent infections capable of causing significant losses during storage, if remedial measures are not taken. However, protection of fresh fruits and vegetables throughout their growing period by use of natural antagonists can be a costly and difficult undertaking. Many preharvest factors and harvest methods influence fresh produce quality during storage and marketing. These variables are very important in evaluating the potential for postharvest loss. This chapter will discuss the importance of some of the major preharvest factors — including weather, cultural practices, and plant nutrition — and their interaction with pest control strategies to affect the postharvest quality of produce. The potential role that biological control strategies may play in these areas to maintain or enhance fresh product quality will also be explored.

If we apply Cook's broad definition of biological control — which is "the use of natural or modified organisms, genes, or gene products to reduce the effects of undesirable organisms (pests) and to favor desirable organisms such as crops, trees, animals, and beneficial insects and microorganisms"[9] — as a starting point for our discussion, we can see that there are many interactions occurring in the environment that can impact on postharvest quality. Let us examine several of the important preharvest factors that may play a significant role in shifting the balance from disease to health, and the way these factors may be manipulated, by employing biological control methods, to enhance product quality.

II. PLANT GERM PLASM

Few opportunities have been explored for the control of specific postharvest diseases by introducing genes for disease resistance. This is in no way meant to minimize the efforts of classical plant breeders to control field diseases that also may adversely affect quality after harvest. Their efforts have greatly increased crop production, in many cases with an associated enhancement of postharvest quality and reduction of storage losses due to decay. Unfortunately, this increase in production has, in some cases, compounded postharvest problems because producers have been forced to develop new transportation and storage strategies to meet increased product yields and expanding markets. The method of preference has been to use chemical control methods when postharvest diseases threaten the quality of fresh produce.[11]

Many important postharvest pathogens are wound pathogens and as such are not amenable to control through breeding for resistance.[4,16,19] Notable exceptions are *Monilinia fructicola* (brown rot of stone fruit) and *Phytophthora infestans* (late blight of potato). These pathogens cause major postharvest losses because of infections which develop during the growing season and spread during storage.

Peach germ plasm with potential resistance to brown rot has been identified in South America; however, the results of disease resistance studies carried out under California growing conditions have not been very promising.[2]

Among potato varieties grown in the U.S. there are genes for resistance to late blight (most notably Kenebec and Sebago), but in no instance is this resistance of high magnitude.[12,17] Little concern has been paid to this problem by plant breeders since the potato industry has depended on fungicide protection to control late blight. However, heightened environmental awareness of fungicide residues is prompting renewed investigation of more resistant germ plasm. There are much better sources of resistant germ plasm in the world than the cultivars grown commercially in the U.S.. However, the difficult problem of introducing genes for resistance into commercial potato varieties is only now becoming a high priority in the U.S..

III. WEATHER

The effects of weather on plant diseases have been the subject of numerous reviews.[3,10] This chapter will explore opportunities to minimize the effects of adverse weather on increasing postharvest diseases and disorders, by employing antagonists and other nonchemical means of disease control before harvest.

Frost injury is a major limiting factor in the production of fresh fruits and vegetables in many temperate growing areas. Globe artichokes are an example of a long season crop for which production is often limited by field frost.[16] Buds freeze at about 29° F. Ice-nucleating bacteria may play a major role in this damage.

Tree fruits, as well as many other small fruits, can also have their production severely limited by late spring frosts which may destroy blossoms and developing fruits. Ice-nucleating bacteria (pseudomonads) have been shown to be a major causal factor of much of this damage. Antagonistic bacteria have been used to displace ice-nucleating pseudomonads in blossoms with excellent results.[14] There are reports in the literature of non-ice nucleating pseudomonads that are also antagonistic to some pathogens responsible for postharvest diseases.[22] It is possible that this dual activity can be exploited to offer a wider base of biocontrol activity against pathogens that cause postharvest diseases. However, a major obstacle to this strategy will be assuring the survival of antagonistic bacteria on the plant surface during the growing season.[5,6]

Investigators may discover additional multiple-use microbial antagonists in nature that can be effectively exploited as broad spectrum biocontrol agents. There is reason for optimism in this area with investigations of bacteria antagonistic to *Erwinia amylovora*, which causes fire blight of apples and pears. Non-ice nucleating pseudomonads and *E. herbicola* strains have been found that act to reduce frost injury and fire blight.[5,14]

Apple scab, caused by *Venturia inaequalis*, is one of the most destructive diseases of apples.[19] This disease is currently controlled by multiple fungicide applications during the growing season. The disease presents a challenge for biocontrol because fruit can become infected throughout the growing season by spores produced on leaves and fruit that were infected early in the growing season. The critical factor for infection is the length of time the leaf surface remains wet enough to allow spore germination. Application of a microbial antagonist to the leaf surface, where it will remain viable during the growing season, may not be feasible, since there is little nutritional substrate available on most leaf and fruit surfaces to sustain most antagonists, much less allow their multiplication, during prolonged periods of desiccation.[21] Alternate periods of high and low humidity would make survival of most bacterial and fungal antagonists unlikely. However, some yeasts appear able to withstand these severe environmental changes.[6] A more practical strategy for biocontrol of *V. inaequalis* may be to attack the pathogen in the leaf and fruit residue on the ground under trees at the end of the growing season. An antagonist that could survive in this environment may have a better chance of limiting apple scab than one forced to survive on leaf and fruit surfaces throughout the growing season. An excellent example of this type of approach has been demonstrated with the control of *Sclerotinia* of lettuce by spray application of *Sporidesmium* spores onto lettuce plants at the end of the growing season, and the subsequent plowing under of the plant debris.[1] There is little doubt that other such interactions occur in nature. Our challenge is to identify them and incorporate those which prove effective into our disease control strategies.

Although there are apple and pear varieties that have high levels of resistance to fruit scab, incorporation of this resistance into commercial varieties has not been a high priority for the industry. Consumer preference may change this position as pressure is brought to bear on the industry to reduce or eliminate fungicide residues on fruit.

Brown rot (*Monilinia fructicola*) is the most destructive market disease of stone fruits, although it primarily infects fruit in the field. There is a significant correlation between the amount of rainfall before harvest and the severity of disease.[4] Some measure of biological control has been achieved using a bacterial antagonist (*Bacillus subtilis*).[18] Colonization of the fruit surface by sufficient bacteria to exclude *M. fructicola* remains an obstacle to effective control of brown rot by this antagonist, as is the case with most other biocontrol agents.[21]

IV. PLANT NUTRITION

Most studies conducted to date on the effects of preharvest nutrition on postharvest diseases and disorders have been related to physiological disorders not directly due to microbial pathogens. In its broadest sense, biological control should also be a major consideration in this area, as we seek natural means to alleviate postharvest disorders and reduce susceptibility to decay. The most promising results have been obtained from research on the effects of calcium on enhanced resistance to postharvest diseases and disorders.[8] Addition of calcium to fruit not only delays senescence, but also reduces susceptibility to a wide variety of postharvest diseases and disorders. Because apples treated with calcium remain firmer than nontreated fruit, senescence is retarded and there is less susceptibility to decay. Unfortunately, calcium content cannot effectively be increased in apples by spray application before harvest because movement is primarily in the xylem and preferentially toward meristematic and transpiring tissues.[7] Potatoes, however, respond to increased levels of calcium in the soil by accumulating it in tubers. A significant correlation between increased calcium in tuber tissue and reduced susceptibility to bacterial soft rot has been demonstrated.[13] Calcium acts to reduce susceptibility to decay by binding to pectins in the cell wall, producing cation bridges between pectic acids, or between pectic acids and other acidic polysaccharides. This hinders the production of host enzymes that cause softening, as well as enzymes produced by pathogens that cause decay.[7]

The application of calcium in combination with microbial antagonists has increased the efficacy of some antagonists.[15] Control of *Botrytis* rot and *Penicillium* rot with reduced concentrations of antagonists was reported in the presence of calcium chloride. Calcium caused a reduction of pathogen spore germination and germ tube elongation. The addition of calcium may be extending a nutrient-limited fungistasis on some postharvest pathogens present on the fruit surface. Under normal conditions, as the fruit matures fungistasis is overcome by the increased availability of nutrients. The addition of calcium to the fruit not only delays senescence by delaying the availability of nutrients, but may also change the ionic environment on the fruit surface enough to reduce the availability of nutrients present. Calcium may also enhance the activity of host defense mechanisms such as phytoalexins and other natural antimicrobial compounds.[20] Unfortunately, induced resistance is not a major factor in most postharvest diseases because of the senescent nature of storage organs.[6]

Suppressive soils have been shown to play an important part in the control of some soilborne plant diseases.[9] Whether this phenomenon can be exploited for the control of postharvest diseases remains to be determined. Limited studies in our laboratory with fluorescent pseudomonads recovered from suppressive soils have not provided positive results. However, this does not prove that these microorganisms may not function in a biocontrol capacity under field conditions.

V. CONCLUSIONS

Although there is little doubt that antagonism exists in nature and that this antagonism may play a major part in the control of postharvest pathogens, the challenge remains for us to identify and exploit instances where antagonists can be of commercial benefit. If antagonistic microorganisms are applied for the control of pathogens, their application must coincide with periods when there is an adequate amount of the pathogen present to sustain the antagonists.[6] While this may be possible in a controlled environment after harvest, it will be much more difficult to achieve prior to harvest. However, since many postharvest pathogens also exist as saprophytes, it may be possible to attack them at vulnerable points in their life cycle. In some cases this may be accomplished by developing resistant germ plasm; in others, by the addition of microbial antagonists or by modifying the environment to reduce host susceptibility. However, it is doubtful that any one strategy will prove to be the secret of success for biocontrol.

Integrated control, allowing for the use of all weapons in the plant pathologist's arsenal, may be our best hope in providing consumers with the best possible produce with the least harmful environmental impact. Holistic plant health care may be the best management tool to help growers achieve these goals.[9] It will allow the management of abiotic and biotic constraints to plant health. Any management system imposed on a crop must be able to identify limiting factors and allow for calculation of cost-benefit ratios where possible, so that a grower does not simply employ a control strategy randomly unless there is a reasonable chance that it will produce a benefit. Modeling systems currently employed for the control of potato late blight and apple scab are good starting points, but they rely heavily on chemical control. Current integrated pest management practices must be modified to minimize the use of synthetic chemicals and promote natural means of disease control. This type of strategy will allow for the conservation and enrichment of beneficial microbial populations, the identification of factors limiting yield, and the maintenance of the growth environment for long-term benefit to the ecosystem.

REFERENCES

1. **Adams, P. B. and Fravel, D. R.**, Economic biological control of *Sclerotinia* lettuce drop by *Sporodesmium sclerotivorum, Phytopathology,* 80, 1120, 1990.
2. **Adaskaveg, J. E., Feliciano, A. J., and Ogawa, J. M.**, Comparative studies of resistance in peach genotypes to *Monilinia fructicola, Phytopathology,* 79, 1183, 1989.
3. **Brooks, C.**, Spoilage of stone fruits on the market, U.S. Department of Agriculture, Circ. 253, 1933.
4. **Bourke, P. M. A.**, Use ,of weather information in the prediction of plant disease epiphytotics, *Annu. Rev. Phytopathol.,* 8, 345, 1970.
5. **Brown, E. W., van der Zwet, T., Bors, R. H., and Janisiewicz, W.**, Identification of a carbohydrate which enhances the growth of a bacterial antagonist against *Erwinia amylovora, Phytopathology,* 82, 718, 1992.
6. **Campbell, R.**, *Biological Control of Microbial Plant Pathogens,* Cambridge Press, 1989.
7. **Clarkson, D. T.**, Calcium transport between tissues and its distribution in the plant, *Plant Cell Environ.,* 7, 449, 1984.
8. **Conway, W. S., Sams, C. E., McGuire, R. G., and Kelman, A.**, Calcium treatment of apples and potatoes to reduce postharvest decay, *Plant Dis.,* 76, 329, 1992.
9. **Cook, R. J.**, Biological control and holistic plant-health care in agriculture, *Am. J. Alternative Agric.,* 3, 51, 1989.
10. **de Villiers, G. D. B., Brown, D. S., Tompkins, R. G., and Green, G. C.**, The Effect of Weather and Climate upon the Keeping Quality of Fruit, Tech Note No. 53, World Meterological Organization, Geneva, 1963.
11. **Eckert, J. W. and Ogawa, J. M.**, The chemical control of postharvest diseases: subtropical and tropical fruits, *Annu. Rev. Phytopathol.,* 23, 421, 1985.
12. **Fry, W. E.**, Management of late blight, in *Advances in Potato Pest Management,* Lashomb, J. H. and Casagrande, R., Eds., Hutchinson Ross Publishing, 1981, 244.
13. **Kelman, A., McGuire, R. G., and Tzeng, K. C.**, Reducing the severity of bacterial soft rot by increasing the concentration of calcium in potato tubers, in *Soilborne Plant Pathogens: Management of Diseases with Macro- and Microelements,* Engelhard, A. W., Ed., American Phytopathological Society, St. Paul, MN, 1989.
14. **Lindow, S. E.**, The role of bacterial ice nucleation in frost injury to plants, *Annu. Rev. Phytopathol.,* 21, 363, 1983.
15. **McLaughlin, R. J.**, A review and current status of research on enhancement of biological control of postharvest diseases of fruit by use of calcium salts with yeasts, in Biological Control Postharvest Dis. Fruits Vegetables, Workshop Proc., Wilson, C. and Chalutz, E., Eds., U.S. Department of Agriculture, ARS-92, 1991, 184.
16. **Moline, H. E. and Lipton, W. J.**, Market diseases of beets, chicory, endive, escarole, globe artichokes, lettuce, rhubarb, spinach, and sweetpotatoes, Agricultural Handbook No. 155, U.S. Department of Agriculture, 1987.
17. **O'Brien, M. J. and Rich, A. E.**, Potato Diseases. Agricultural Handbook No. 474, U.S. Department of Agriculture, 1976, 25.
18. **Pusey, P. L. and Wilson, C. L.**, Postharvest biological control of stone fruit brown rot by *Bacillus subtilis, Plant Dis.,* 68, 753, 1984.
19. **Rose, D. H., McColloch, L. P., and Fisher, D. F.**, Market Diseases of Fruits and Vegetables — Apples, Pears, Quinces. U.S. Department of Agriculture Miscellaneous Pubublication 168, 1951.

20. **Schonbeck, F. and Dehne, H. W.,** Use of microbial metabolites inducing resistance against plant pathogens, in *Microbiology of the Phyllosphere,* Fokkema, N. J. and Van den Heuvel, J., Eds., Cambridge University Press, 1986, 363.

21. **Spurr, H.,** Managing ephiphytic microflora for biological control, in Biological Control Postharvest Dis. Fruits Vegetables, Workshop Proc., Wilson, C. and Chalutz, E., Eds., Agricultural Research Service, U.S. Department of Agriculture, ARS-92, 1991, 3.

22. **Whipps, J. M.,** Use of microorganisms for biological control of vegetable diseases, *Aspects Appl. Biol.,* 12, 75, 1986.

Chapter 6

Mode of Action of Biocontrol Agents of Postharvest Diseases

Samir Droby and Edo Chalutz

CONTENTS

I. INTRODUCTION

Biological control of postharvest diseases of fruits and vegetables is a relatively new research area.[1] The intensive efforts that have been made in recent years to screen for, isolate, and test the efficacy of biocontrol agents under laboratory and semicommercial conditions have resulted in many cases of microorganisms reported to protect fruits from infection.

Our knowledge on the mode of action of most postharvest biocontrol agents is meager. It has mostly followed concepts of mode of action based on traditional screening procedures. These would favor the selection of antagonists which act by direct interactions or by the production and secretion to the growth medium of substances toxic to the pathogens (antibiotics). Screening programs based on this concept are likely to disregard potential antagonists among the epiphytic microflora with other modes of action.

The lack of thorough knowledge on the modes of action of postharvest biocontrol agents, other than antibiosis, may be attributed to our limited understanding of the interactions taking place between the host, the pathogen, and the antagonist at the site of infection. Yet, information on the mechanisms of antagonism is crucial for developing successful postharvest biocontrol strategies. Such information is essential for: (1) optimization of the method and timing of application of the antagonist, (2) developing appropriate formulations to enhance antagonist efficacy, (3) developing a rationale for selecting more effective antagonists, and (4) registration of biocontrol agents for commercial use.

In general, the mode of action of many antagonists of postharvest diseases is poorly understood. In the absence of the production of antibiotics, it appears that the mode of action comprises a complex mechanism which could involve one or several of the following processes: nutrient competition, site exclusion, induced host resistance, and direct interaction between the antagonist and the pathogen.

Considerable difficulties are encountered in the elucidation of the mode of action of postharvest biocontrol agents because of the problematic interpretation of information obtained from *in vitro* studies. While it may be relatively easy to study features of the antagonist-pathogen interactions in culture, it is much more difficult to prove the

0-8493-4567-7/94/$0.00+$.50

involvement of a mechanism taking place at the site of action, namely, at the wound or nonwounded fruit tissue. Hence, most conclusions regarding modes of action are often based on indirect evidence.

In 1988 we summarized the information available on the modes of action of postharvest biocontrol agents.[2] In this chapter we review the information available to date, pointing out topics on which information is lacking and difficulties encountered in the elucidation of modes of action. Areas of future research on mode of action of postharvest biocontrol agents are also proposed.

II. MODES OF ACTION

A. ANTIBIOSIS

Secretion of antibiotic substances is a common phenomenon in nature. A number of bacterial antagonists have been reported to produce antibiotics *in vitro* and may have a role in protecting commodities against diseases before and after harvest. In screening tests for naturally occurring microbial antagonists of postharvest diseases of citrus and deciduous fruit, Wilson and Chalutz[3] have screened a large number isolates of the epiphytic microbial population obtained from the surface of fruits. Many of these isolates showed inhibition against several postharvest pathogens in culture. However, production of antibiotics in culture media may not be necessarily indicative of their production at the site of action on the fruit surface. In fact, some antibiotics have been shown to be produced in culture only.[4]

Gutter and Littauer[5] were among the first to report that the bacterium *Bacillus subtilis* possesses *in vitro* inhibitory activity against a wide rage of pathogens of fruits. More recently, Pusey and Wilson,[6] Singh and Deverall,[7] Korsten et al.,[8] and Utkhede and Sholberg[9] have studied the use of *B. subtilis* for the control of important postharvest diseases of fruits and vegetables. Pusey and Wilson[6] reported that the B-3 strain of *B. subtilis* was effective in the control of postharvest rot of stone fruit under laboratory as well as under simulated commercial conditions.[10-12] In initial screening tests, a number of bacteria produced zones of inhibition when cocultured on solid media with the stone fruit pathogen *Monilinia fructicola*, but had little or no effect against the fungus when tested on fruit. The B-3 strain, however, proved effective both in culture and on fruit. Cell-free filtrates of the bacterium protected fruit from the pathogen. The biologically active materials were isolated and identified as iturin peptides.[13,14] These antibiotics are active against very few bacteria but have a wide antifungal spectrum, have low toxicity and lack allergenic properties, and were of potential value in topical treatment of fungal skin diseases.[15] However, the use of *B. subtilis* on harvested commodities has not met with a favorable response because it was suspected of being associated with food poisoning in the U.K., Australia, and New Zealand,[16] although it was never proved to be the causal agent of the poisoning.

Several other reports have associated *B. subtilis* and other fungal and bacterial antagonists such as *B. pumilus*, *Myrothecium roridum,* and *M. verrucaria* with the inhibition of *Penicillium digitatum* in lemon fruit wounds through the secretion of antibiotics.[7,17,18]

Colyer and Mount[19] reported that a strain of *Pseudomonas putida* protected potato tubers against postharvest soft rot diseases caused by *Erwinia* spp. when applied as a postharvest treatment. Potato tubers treated after harvest with an antibiotic producing isolate of *P. putida* showed less soft rot than did the nontreated control tubers. Treatment of the tubers with an antibiotic negative strain gave intermediate reduction of soft rot. Based on these findings it was assumed that biosis may be responsible for the inability of the antibiotic negative bacterium to reduce the soft rot to the same level as did the antibiotic positive strain.

The bacterium *P. cepacia* was shown by Janisiewicz and Roitman[20] to be effective in the control of postharvest diseases of apples and pears. This bacterium produced pyrrolnitrin, a powerful antifungal compound. The compound inhibited *Botrytis cinerea* and *P. expansum* in culture at concentrations below 1 mg/ml. Apples and pears treated with pyrrolnitrin generally showed reduced development of decay from the blue mold and gray mold fungi. In addition, four other chlorinated phenylpyrrole derivatives were isolated from *P. cepacia* which possesses antifungal activity against blue and gray mold of apples and pears.[21,22] The role of pyrrolnitrin in controlling the green mold disease of citrus fruit, however, has been recently questioned by Smilanick and Denis-Arrue.[23] In their work, a pyrrolnitrin-resistant isolate of *P. digitatum* was controlled by *P. cepacia* applied to lemon fruit wounds challenged with the pathogen spores. These results suggested that other factors, such as nutrient depletion or space exclusion, may be involved in the inhibition mechanism of *P. cepacia*.

Another question related to antibiotic action in biocontrol is whether production of the pathogen inhibitory substance in culture is indicative of its involvement on the plant. Kraus and Loper[24] recently reported the lack of evidence for the involvement of an antibiotic substance produced *in vitro*, in the *in vivo* biocontrol activity of *Pseudomonas fluorescens* against *Pythium* damping-off of cucumber.

The question of whether antibiotic-producing antagonists should be screened for and used for the control of postharvest diseases of fruits and vegetables is still an open one. On the one hand, the introduction of antibiotic-producing biocontrol agents to our food may have an adverse effect on human resistance to antibiotics, and therefore such biocontrol agents may not be desirable or allowed for use. In addition, such antagonists are not preferable since pathogens are likely to develop resistance to their toxic metabolites much faster than they are to antagonists with other modes of action, possibly by a single mutation. Thus, their efficacy may be lost as rapidly as that of some synthetic fungicides. On the other hand, antibiotic-producing antagonists may be more effective biocontrol agents than antagonists with other mechanisms, since they may better protect infections which occur prior to their application due to the penetration of the materials produced. Also, it can be argued that through history, the fermentation industry has been using organisms that produce antibiotic substances that are inhibitory to spoilage organisms, a practice that has been considered safe. Therefore, it is realistic to assume that microorganisms used as biocontrol agents and produce antibiotic substances at the level found in fermented foods may not necessarily be harmful. The potential development of pathogen resistance, therefore, is a major consideration against the use of antibiotics in plants. However, natamycin is an example of an antifungal compound widely used for food preservation, to which very little resistance is known to have been developed.[25] It is likely, therefore, that other antibiotics with this attribute also exist.

In order to determine whether the use of an antibiotic-producing antagonist is desirable, several factors have to be weighed. First, knowledge on the involvement of antibiotics in the mode of action should be generated. Then, information on the possible involvement of other factors in the mode of action of known antibiotic-producing antagonists needs be evaluated. The final decision should take into consideration all the factors mentioned above, in addition to the human safety consideration.

B. COMPETITION FOR NUTRIENTS AND SPACE

The term competition has been defined by Wicklow[26] as niche overlap, where there is simultaneous demand on the same resources by two or more microbial populations. Bacteria and yeast, partly because of their large surface-to-volume ratio, are able to take up nutrients from dilute solutions more rapidly and in greater quantity than the germ tubes of filamentous fungal pathogens.[27,28] Indeed, competition for nutrients on the phylloplane is a widespread and well-documented phenomenon.[29] For example, epiphytic bacteria have been shown to reduce germination of conidia of *B. cinerea*, possibly by removing amino acids from amino acid and glucose mixtures more quickly than did the conidia of the pathogen.[27,30]

Most postharvest infections of fruits and vegetables occur through surface wounds inflicted during harvest and subsequent handling. To control wound pathogens biologically, the antagonist must normally be present at the wound site where the antagonist-pathogen interactions occur. An effective antagonist should, therefore, possess special features to be a successful competitor at the wound site: it should be better adapted than the pathogen to extreme environmental and nutritional conditions; grow rapidly at the wound site; be an effective utilizer of nutrients at low concentrations; and survive and develop better than the pathogen on the surface of the commodity and at the infection site under extreme temperature, pH, and osmotic conditions.

Indeed, recent studies on biological control of postharvest diseases of fruits and vegetables have reported the use of microorganisms that multiply rapidly, colonize the wound, and successfully compete for nutrients and space. For example, the yeasts *Pichia guilliermondii* (formerly identified as *Debaryomyces hansenii*), *Cryptococcus laurentii*, *Aureobasidium pullulans*, Candida spp., and the bacteria *Enterobacter cloacae* and *Pseudomonas cepacia* have all been reported to rapidly and extensively colonize the wound site.[2,31-36]

Wound competence under different environmental conditions may be a key factor for the evaluation of microbial agents. *P. cepacia,* for example, was reported to survive and increase rapidly in number in wounds of lemon fruit[23] and apple and pear fruits.[20] On nonwounded fruit surfaces, however, *P. cepacia* survived poorly. On the other hand, Droby et al.[31] reported that the US-7 isolate of *Pichia guilliermondii* survived for long periods both at the wound site and the nonwounded fruit surface. Rapid colonization of apple fruit wounds by *C. laurentii* at temperatures ranging from 0 to 20° C has been reported by Roberts,[35] who also reported that *C. laurentii* and *C. flavus* rapidly colonize wounds of apple and pear fruits under ambient and controlled atmospheric conditions (1.5% CO_2, 2.0% CO_2) unfavorable to the development of the pathogen.

In our studies we observed[31] that the US-7 isolate of *P. guilliermondii* multiplied very rapidly at the wound site under a wide range of temperature, humidity, and nutritional conditions and that it may increase in numbers, by 1 to 2 orders of magnitude within 24 h while at the end of this incubation period, the pathogen spores had just started to germinate and grow. The growth rate of the yeast at the wound site at different temperatures and its persistency in the tissue is shown in Figure 1A. The ability of the antagonist cells to rapidly increase in number at relatively low temperatures, compared with the pathogen, explains the observed increase in efficacy against the green mold of grapefruit when efficacy is compared at decreasing storage temperatures (Figure 1B). At the relatively high temperature of 25° C, the low yeast cell count used in these tests was only partially effective, reducing green mold infections by only 70 to 10% of control. At lower storage temperatures, not only was infection delayed, as expected by the reduced temperature, but also the efficiency of the antagonist has actually increased: at 6° C the incidence of infection was reduced by 100 to 45% of control.

Several lines of evidence supported the assumption that the inhibition of the pathogen development by the antagonist involves competition for nutrients.[31] Such competition was demonstrated by *P. guilliermondii* in culture, when both the antagonist and the pathogen were cocultured in a minimal synthetic medium or in wound leachate solutions.[31] The efficacy of the yeast could be markedly reduced by the addition of nutrients to the spore suspension used for inoculation. Similarly, *Enterobacter cloacae*, a bacterium antagonist, inhibited germination of *Rhizopus stolonifer* spores through nutrient competition.[32] In both studies, indirect evidence was provided to demonstrate the role of competition for nutrients as a mode of action of these two antagonists: (1) inhibition of spore germination or growth of the pathogen during coculturing with the antagonist was demonstrated; (2) inhibition of the pathogen was dependent on the concentration of the antagonist propagules, and (3) partial or complete reversal of the inhibition could be achieved by the addition of exogenous nutrients.

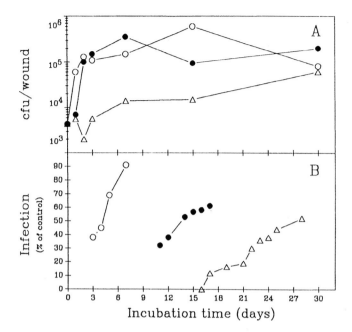

Figure 1 Growth rate of *Pichia guilliermondii* on surface wounds of grapefruit (**A**) and its biocontrol activity against *Penicillium digitatum* (**B**) at storage temperatures of 25° C (open circles), 11° C (closed circles), and 6° C (triangles).

In most reports on biological control of postharvest diseases of fruits and vegetables, a quantitative relationship has been demonstrated between the antagonist concentration and the biocontrol effectiveness. Thus, a delicate balance apparently exists at the wound site between the numbers of antagonist and pathogen propagules, which affects the outcome of the interaction and determines whether the wound becomes the site of infection. Manipulation of the initial concentration of the antagonist cells and/or the fungal spores clearly affects infection. On the other hand, we have shown that the number of antagonist cells at the wound site will not always determine its efficacy.[37] Our data suggested that active multiplication and growth of the US-7 yeast cells were required for the yeast to exhibit its biocontrol activity. This was demonstrated by using a mutant of *P. guilliermondii* which lost its biocontrol activity against *Penicillium digitatum* on grapefruit and against *B. cinerea* on apples, even when applied to the wound at concentrations as high as 10[10] cells per milliliter. At the wound sites, the cell population of this mutant remained constant during the incubation period, while that of the wild type increased 60- to 100-fold, within 24 h. Failure of the mutant to inhibit spore germination of the pathogen in culture on a minimal salt medium suggested that this mutant lost its ability to utilize some nutrients and grow in culture as well. This could be the reason for its nonefficacy.[38]

While competition for nutrients is likely a widespread phenomenon in the interaction between microorganisms on the phylloplane, the demonstration of nutrient competition as a mechanism of antagonism assumes that the pathogenic fungus depends on an external source of nutrients for germination and penetration into the host tissue. This assumption is very difficult to demonstrate.

It would be worthwhile to mention here that while in macroecology it has long been axiomatic that competition for nutrients plays the central role,[39,40] Andrews and Harris[41] and Andrews[42] more recently stated that except under certain conditions competition is probably not a major force on the phylloplane, or at least that evidence for its role under natural conditions is either lacking or largely indirect.

In summary, rapid growth and extensive colonization of the wound site by the antagonist are important features of many reported postharvest biocontrol agents. Active growth of antagonist cells, presumably depletes nutrients or denies available space from the pathogen and thus reduces its growth rate and the incidence of infection. For effective levels of biocontrol activity, the antagonist cells must probably reach a critical number thus avoiding the establishment of the pathogen in the fruit or vegetable tissue. In most cases, therefore, application of the biocontrol agents after the establishment of the pathogen in the tissue resulted in a much reduced efficacy of the biocontrol agent.

C. INDUCTION OF RESISTANCE MECHANISMS IN THE HOST

Induced resistance has been recognized as an important and manipulatable form of resistance in vegetative plant tissues.[42] Similar mechanisms of resistance may operate in harvested fruit and vegetable tissues. Indeed, some reports indicated[32,37] that certain postharvest biocontrol agents may interact with the host tissue, in particular with wounded surfaces, leading to enhanced wound-healing processes. However, documented direct evidence to support this possible mode of action is lacking.

As indicated above, several non-antibiotic-producing yeast antagonists of wound pathogens are most effective when their application occurs prior to inoculation by the pathogen. Application of the antagonist cells after inoculation resulted in decreased efficacy. Chalutz and Wilson[44] reported that the longer the time elapsing between infection and antagonist application, the less was the antagonist efficacy. At 25° C, only 30% reduction of the incidence of disease was observed when the antagonist was applied 7 h after inoculation and no reduction was evident when antagonist application was done at 24 h after inoculation, compared with more than 90% reduction in disease incidence when the antagonist and pathogen were applied simultaneously. While this trend has been demonstrated in laboratory studies, in larger scale semicommercial tests reduction of yeast efficacy due to its application after inoculation or natural infection was lower than exhibited in the laboratory tests. These observations suggested the possibility that application of the yeast cells may induce resistance processes in the peel tissue. To test this hypothesis, we examined the production of ethylene by yeast-treated tissues; when cell suspensions of the US-7 yeast antagonist were placed on surface wounds of grapefruit, pomelo, table grapes, or carrot root tissue, enhanced ethylene production was evident in all tissues (Figures 2A, B, C, and D). In carrot root disks, which were used as a model system, the application of the yeast antagonist resulted not only in enhancment of ethylene production but also in accumulation of phenols or phenollike materials which exhibited absorption at 280 nm (Figure 3). When cultured *in vitro*, the yeast cells themselves did not produce ethylene.[38,44] Involvement of ethylene in the induction of resistance processes in grapefruit and carrot roots was demonstrated in the past,[45,46] possibly through the induction of the activity of phenylalanine ammonia-lyase (PAL), an enzyme which catalyzes the branch point step reactions of the shikimic acid pathway; this leads to the synthesis of phenols, phytoalexins, and lignins, all associated with induced resistance processes.[43,47,48] In citrus fruit, ethylene production and PAL activity was induced following application to peel disks of an effective yeast antagonist,[38] while exogenously applied ethylene to the disks or to whole grapefruits induced resistance to *P. digitatum* infections (Table 1). Thus the induction of ethylene production during interaction of the antagonist with the tissue suggests the involvement of host resistance mechanisms in the yeast action. The nature of this mechanism is yet to be elucidated.

Induced resistance of plants by nonpathogenic microorganisms has been studied in many plant systems.[43] Potato tuber tissue was shown by Henderson and Friend,[49] Friend,[50,51]

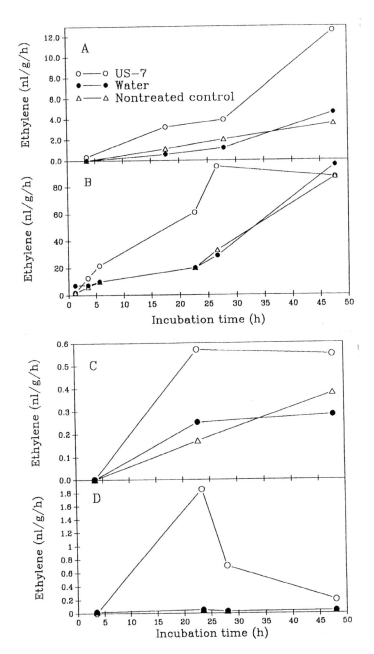

Figure 2 Induction of ethylene production by *Pichia guilliermondii* in peel disks of grapefruit (**A**), pomelo fruit (**B**), grape berries (**C**), and carrot root disks (**D**).

and by others[52] to deposit ligninlike material more rapidly in response to incompatible races of *Phytophthora infestans* than to compatible races. Hammerschmidt[53] has also demonstrated the use of nonpathogenic *Cladosporium cucumerinum* for induction of lignin in potato tuber tissue. The use of nonpathogenic microorganisms to induce resistance in the harvested commodity has not been fully explored. However, it should be

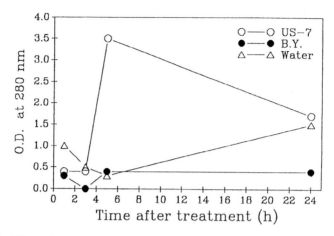

Figure 3 Effect of carrot root disk treatment with an effective yeast biocontrol agent (US-7) and a noneffective yeast (B.Y.) on the accumulation of phenols or phenollike materials.

Table 1 **Ethylene-induced resistance of grapefruit to *Penicillium digitatum* infections**

| | *Penicillium digitatum*[a] (spores per milliliter) | |
	10^4	10^5
Air control	20	39
Ethylene[b]	0	8

Note: Freshly harvested grapefruits were exposed to 20 ppm ethylene for 48 h and then inoculated with a spore suspension of *Penicillium digitatum* through artificial peel wounds. Percent infection was determined after 5 d of incubation at 20 C.

[a] % Infection; [b] 20 ppm.

possible to screen for epiphytic microbial components that may induce resistance processes in the wound to prevent the entry of postharvest pathogens.

D. DIRECT INTERACTIONS WITH THE PATHOGEN

Direct parasitism of the antagonist to the pathogen propagules has been reported to play a role in biological control against soilborne[54] and foliar diseases.[55] A detailed study in this regard has been conducted with *Trichoderma*[54,56-58] which has been reported to antagonize pathogens by direct parasitism. In the postharvest arena, however, very little information is available on biological control agents that directly parasitize pathogens. Wisniewski et al.[59] have shown that the yeast antagonist *Pichia guilliermondii*, when cocultured with *Botrytis cinerea*, appears to strongly attach to *B. cinerea* hyphae (Figure 4). This attachment was blocked when the yeast cells or the pathogen hyphae were exposed to compounds that affect protein integrity, or when respiration was inhibited. A lectin-type binding was suggested in this attachment. In addition, *P. guilliermondii* was found to exhibit high levels of β-1,3-glucanase activity when cultured on various carbon

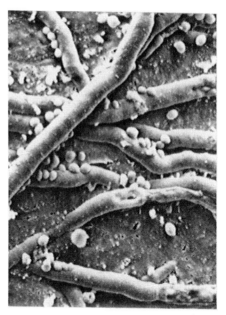

Figure 4 Attachment of cells of *Pichia guilliermondii* to the mycelium of *Botrytis cinerea* and its degradation.

sources or on cell walls of several fungal pathogens.[59,60] The fastidious attachment of this yeast antagonist to fungal cell wall would enhance the effectiveness of any cell wall hydrolases secreted by the yeast to the extracellular matrix. When yeast cells were dislodged from the hyphae, concave appearance of hyphal surface and partial degradation of *B. cinerea* cell wall was also observed at the attachment sites (Figure 4). Thus, the firm attachment of the yeast cell along with the production of hydrolases may be responsible for the observed degradation of fungal cell wall. Production of glucanase may also enhance the ability of the yeast to adhere to fungal hyphae, as suggested for *Candida albicans.*[61]

In addition, we have recently observed that an extracellular polysaccharide extracted from the surface of cells of the yeast *P. guilliermondii* exhibited antifungal activity against postharvest pathogens,[60] indicating its possible involvement in the yeast's mode of action. The processes involved in this interaction are currently under investigation.

Thus, biocontrol activity of *P. guilliermondii* — and possibly other yeast biocontrol agents — not only may be dependent on its ability to rapidly colonize the wound site and compete for nutrients, but also may depend on its ability to attach firmly to hyphae of the pathogen and to produce cell wall degrading enzymes.

III. FUTURE RESEARCH NEEDS

Although antibiotic substances have been reported to be involved in the mode of action of some bacterial antagonists of postharvest diseases, conclusive evidence for their involvement in biocontrol activity on the fruit is lacking. Development of assays for antibiotic substance production and presence at the wound site, and the use of antibiotic-deficient mutants and/or antibiotic-resistant pathogens, would help elucidate the role of antibiotics in the mode of action of biocontrol agents.

Special attention should be given to study the site of interaction between antagonist and pathogen on the surface of the commodity. Of particular interest is the chemical composition of wound leachates and presence of substances that may favor growth of the antagonist or pathogen and thus affect the dynamics of the interactions. Also in terms of antagonist-host interactions, the mechanisms of microbial adhesion, spatial patterns of colonization, and nature of resources on fruit surfaces and particularly in wounds need to be thoroughly studied.

Our understanding of defense mechanisms in the harvested commodity is very limited. Once this knowledge is expanded to cover the complex reactions of induced resistance and the chemical and physical elicitors that increase resistance in harvested fruits, the possibility of selecting and using effective microbial antagonists that induce resistance will be much enhanced.

Understanding the wound competence of microbial antagonists at the molecular level is required. Now, when tools of molecular biology and genetic engineering are available, it should be feasible to determine the gene(s) involved in rapid growth of the antagonist at the wound environment and other factors facilitating the effective performance of the antagonist. Molecular biology techniques will also improve our ability to investigate the mode of action and enable us to conduct more accurate ecological impact and population dynamics studies.

A complex interaction at the wound site may be involved and eventually determine the efficacy of the antagonist. In this regard, several factors have to be considered: type of the wound and plant tissue, fruit maturity, chemical composition and pH value of components of wound leachate, wound healing processes, and interactions of antagonist and pathogen with the host tissue and with other microorganisms developing at the wound. Increasing our knowledge in these areas will, undoubtedly, contribute to a better evaluation of the nature of the interactions taking place in the wounded and nonwounded fruit surface.

IV. CONCLUDING REMARKS

Elucidation of the mode of action of postharvest biocontrol agents in relation to the etiology of the disease antagonized is essential for successful use, and is directly relevant to the implementation of this new technology. As research on postharvest biological control continues, our knowledge on mode of action is likely to increase.

The assumption that a mode of antagonism as determined *in vitro* also occurs *in vivo*, might be inaccurate. The procedures and criteria for finding and selecting antagonists to control postharvest diseases have not yet been adequately defined. Knowledge on mechanisms of action would be used also, whenever possible, to devise methods of selecting more active strains of antagonists.

Another factor that might alter biocontrol efficacy is the addition of formulation ingredients. Therefore, these should be carefully selected based on the specific modes of interaction between the host, the pathogen, and the antagonist.

The picture emerging from our efforts to elucidate the mode of action of postharvest biocontrol agents shows no obvious indication that one major mechanism is dominant in the biocontrol activity of most antagonists. Thus, it is likely that multiple interactions between antagonist, host, pathogen, and other components of fruit natural epiphytic microflora take place by different modes of action. Development of a simulation model to guide the introduction of biocontrol agents on fruit surfaces before or after harvest could, therefore, be of much help.

More complete knowledge of the modes of action of biocontrol agents of postharvest diseases will enable us to use tools of genetic engineering and molecular biology techniques to improve properties and subsequently enhance efficacy. However, an effective biocontrol agent may not be developed until the dynamics of the interactions of antagonist, pathogen, plant, epiphytic microflora, and environmental factors are more fully understood.

REFERENCES

1. **Wisniewski, M. E. and Wilson, C. L.,** Biological control of postharvest diseases of fruits and vegetables: recent advances, *HortScience,* 27, 94, 1992.
2. **Wilson, C. L. and Wisniewski, M. E.,** Biological control of postharvest diseases of fruits and vegetables: an emerging technology, *Annu. Rev. Phytopathol.,* 27, 425, 1989.
3. **Wilson, C. L. and Chalutz, E.,** Postharvest biocontrol of *Penicillium* rots of citrus with antagonistic yeasts and bacteria, *Sci. Hortic.,* 40, 105, 1989.
4. **Weinberg, E. D.,** Biosynthesis of secondary metabolites: role of trace metals, *Adv. Microbiol. Physiol.,* 4, 1, 1970.
5. **Gutter, Y. and Littauer, F.,** Antaginistic action of *Bacillus subtilis* against citrus fruit pathogens, *Bull. Res. Coun. Isr.,* 3, 192, 1953.
6. **Pusey, P. L. and Wilson, C. L.,** Postharvest biological control of stone fruit brown rot by *Bacillus subtilis, Plant Dis.,* 68, 753, 1984.
7. **Singh, N. and Deverall, B. J.,** *Bacillus subtilis* as a control agent against fungal pathogens of citrus fruit, *Trans. Br. Mycol. Soc.,* 83, 487, 1984.
8. **Korsten, L., Villiers, E. E., Jager, E. S., Cook, N., and Kotze, J. M.,** Biological control of avocado postharvest diseases, *S. Afr. Avocado Growers' Assoc. Yearb.,* 14, 57, 1991.
9. **Utkhede, R. S. and Sholberg, P. L.,** *In vitro* inhibition of plant pathogens by *Bacillus subtilis* and *Enterobacter aerogenes* and *in vivo* control of two postharvest cherry diseases, *Can. J. Microbiol.,* 32, 963, 1986.
10. **Wilson, C. L. and Pusey, P. L.,** Potential for biological control of postharvest plant diseases, *Plant Dis.,* 69, 375, 1985.
11. **Pusey, P. L., Wilson, C. L., Hotchkiss, M. W., and Franklin, J. D.,** Compatibility of *Bacillus subtilis* for postharvest control of peach brown rot with commercial fruit waxes, dichloran and cold-storage conditions, *Plant Dis.,* 70, 578, 1986.
12. **Pusey, P. L., Hotchhkiss, M. W., Dulmage, H. T., Baumgardner, R. H., Zehr, E. I., Reilly, C. C., and Wilson, C. L.,** Pilot tests for commercial production and application of *Bacillus subtilis* (B-3) for postharvest control of peach brown rot, *Plant Dis.,* 72, 622, 1988.
13. **Gueldner, R. C., Reilly, C. C., Pusey, P. L., Costello, C. E., Arrendale, R. F., Himmelsbach, D. S., Crumley, F.G., and Culter, H. G.,** Isolation and identification of iturins as antifungal peptides in biological control of peach brown rot with *Bacillus subtilis, J. Agric. Food. Chem.,* 36, 366, 1988.
14. **McKeen, C. D., Reilly, C. C., and Pusey, P. L.,** Production and partial characterization of antifungal substances against *Monilinia fructicola* from *Bacillus subtilis, Phytopathology,* 76, 136, 1986.
15. **Delcambe, L., Peypoux, F., Guinand, M., and Michael, G.,** Structure of iturin and iturin-like substances, *Rev. Ferment. Ind. Aliment.,* 31, 147, 1976.
16. **Kramer, J. M. and Gilbert, R. J.,** *Bacillus cereus* and other *Bacillus:* chemistry, biogenesis, and possible functions, *Bacteriol. Rev.,* 41, 449, 1989.
17. **Appel, D. J., Gees, R., and Coffey, M. D.,** Biological control of the postharvest pathogen *Penicillium digitatum* on Eureka lemons, *Phytopathology,* 78, 1593, 1988.
18. **Huang, Y., Wild, B. L., and Morris, S. C.,** Postharvest biological control of *Penicillium digitatum* decay on citrus fruit by *Bacillus pumilus, Ann. Appl. Biol.,* 120, 367, 1992.
19. **Colyer, P. D. and Mount, M. S.,** Bacterization of potatoes with *Pseudomonas putida* and its influence on postharvest soft rot disease, *Plant Dis.,* 68, 703, 1984.
20. **Janisiewicz, W. J., and Roitman, J.,** Biological control of blue and gray mold on apple and pear with *Pseudomonas cepacia, Phytopathology,* 78, 1697, 1988.

74

21. **Roitman, J. N., Mahoney, N. E., and Janisiewicz, W. J.,** Production and composition of phenylpyrrole metabolites production by *Pseudomonas cepacia, Appl. Microbiol. Biotechnol.,* 34, 381, 1990.

22. **Janisiewicz, W. J., Yourman, L., Roitman, J., and Mahoney, N.,** Postharvest control of blue mold and gray mold of apples and pears by dip treatments with pyrolnitrin, a metabolite of *Pseudomonas cepacia, Plant Dis.,* 75, 490, 1991.

23. **Smilanick, J. L. and Denis-Arrue, R.,** Control of green mold of lemons with *Pseudomonas* species, *Plant Dis.,* 76, 481, 1992.

24. **Kraus, J. and Loper, J. E.,** Lack of evidence for the role of antifungal metabolite production by *Pseudomonas fluorescens* PF-5 in biological control of *Pythium* damping-off of cucumber, *Phytopathology,* 82, 264, 1992.

25. **Jay, J. M.,** Antibiotics as food preservatives, in *Food Microbiol.,* 8, 117, 1983.

26. **Wicklow, D. T.,** Interference competition and the organization of fungal communities, in *The Fungal Community: Its Organization and Role in the Ecosystem,* Wicklow, D. T. and Caroll, G. C., Eds., Marcel Dekker, New York, 1981, 351.

27. **Bordie, I. D. S. and Blakeman, J. P.,** Competition for exogenous substrates *in vitro* by leaf surface microorganisms and germinating conidia of *Botrytis cinerea, Physiol. Plant Pathol.,* 9, 227, 1976.

28. **Fokkema, N. J.,** Fungal leaf saprophytes, beneficial or detrimental?, in *Microbial Ecology of the Phylloplane,* Blakeman, J. P., Ed., Academic Press, New York, 1981, 433.

29. **Blakeman, J. P. and Bordie, I. D. S.,** Competition for nutrients between epiphytic microorganisms and germination of spores of plant pathogens on beetroot leaves, *Physiol. Plant Pathol.,* 10, 29, 1977.

30. **Blakeman, J. P. and Fraser, A. K.,** Inhibition of *Botrytis cinerea* spores by bacteria on the surface of chrysanthemum leaves, *Physiol. Plant Pathol.,* 1, 45, 1971.

31. **Droby, S., Chalutz, E., Wilson, C. L., and Wisniewski, M.,** Characterization of the biocontrol activity of *Debaryomyces hansenii* in the control of *Penicillium digitatum* on grapefruit, *Can. J. Microbiol.,* 35, 794, 1989.

32. **Wisniewski, M., Wilson, C. L., and Hershberger, W.,** Characterization of inhibition of *Rhizopus stolonifer* germination and growth by *Enterobacter cloacae, Can. J. Bot.,* 67, 2317, 1989.

33. **Roberts, R. G.,** Postharvest biological control of gray mold of apple by *Cryptococcus laurentii, Phytopathology,* 80, 526, 1990.

34. **Roberts, R. G.,** Biological control of mucor rot of pear by *Cryptococcus laurentii, C. flavus,* and *C. albidus, Phytopathology,* 80, 1051, 1990.

35. **Roberts, R. G.,** Characterization of postharvest biological control of deciduous fruit diseases by *Cryptococcus* spp., in *Proc. Int. Workshop Biological Control of Postharvest Dis. Fruits Vegetables,* Wilson, C. L. and Chalutz, E., Eds., Agricultural Research Services, U.S. Department of Agriculture, ARS-92, 1991, 32.

36. **Gullino, M. L., Aloi, C., Palitto, M., Benzi, D., and Garibaldi, A.,** Attempts at biological control of postharvest diseases of apple, *Med. Fac. Landbouww. Rijksuiv. Genet.,* 56, 195, 1991.

37. **Droby, S., Chalutz, E., Cohen, L., Weiss, B., Wilson, C. L., and Wisniewski, M.,** Nutrient competition as a mode of action of postharvest biocontrol agents, in *Proc. Int. Workshop Biological Control of Postharvest Dis. Fruits Vegetables,* Wilson, C. L. and Chalutz, E., Eds., Agricultural Research Service, U.S. Department of Agriculture, ARS-92, 1991, 142.

38. **Droby, S., Chalutz, E., and Wilson, C. L.,** Antagonistic microorganisms as biocontrol agents of postharvest diseases of fruits and vegetables, *Postharvest News Inf.,* 2, 169, 1991.

39. **Miller, R. S.,** Patterns and process in competition, *Adv. Ecol. Res.,* 4, 1, 1979.

40. **Pianka, E. R.,** *Evolutionary Ecology,* Harper & Row, New York, 1978.
41. **Andrews, J. H. and Harris, R. F.,** r- and K-selection and microbial ecology, in *Advances in Microbial Ecology,* Marshall, K. C., Ed., Plenum Press, New York, 1986, 99.
42. **Andrews, J. H.,** Life history of plant pathogens, *Adv. Plant Pathol.,* 2, 105, 1984.
43. **Kuc', J.,** Induced immunity to plant disease, *Bioscience,* 32, 854, 1982.
44. **Chalutz, E. and Wilson, C. L.,** Postharvest biocontrol of green and blue mold and sour rot of citrus fruit by *Debaryomyces hansenii, Plant Dis.,* 74, 134, 1990.
45. **Lisker, N., Cohen, L., Chalutz, E., and Fuchs, Y.,** Fungal infection suppress ethylene-induced phenylalanine ammonia-lyase activity in grapefruits, *Physiol. Plant Pathol.,* 22, 331, 1983.
46. **Chalutz, E., DeVay, J. E., and Maxie, E.C.,** Ethylene-induced isocoumarin formation in carrot root tissue, *Plant Physiol.,* 44, 235, 1969.
47. **Halbrock, K. and Grisebach, H.,** Enzymatic control of the biosynthesis of lignin and flavonoids, *Annu Rev. Plant Physiol.,* 30, 105, 1979.
48. **Halbrock, K. and Scheel, D.,** Biochemical responses of plant to pathogens, in *Innovative Approaches to Plant Disease Control,* Chet, I., Ed., John Wiley & Sons, New York, 1987, 229.
49. **Henderson, S. J. and Friend, J.,** Increase in PAL and lignin-like compounds as race-specific resistance responses of potato tubers to *Phytophthora infestans, Phytopathol. Z.,* 94, 323, 1979.
50. **Friend, J.,** Plant phenolics, lignification and plant diseases, *Prog. Phytochem.,* 7, 197, 1981.
51. **Friend, J.,** Phenolics and resistance — some speculations, in *Active Defense Mechanisms in Plants,* Wood, R.K.S., Ed., Plenum Press, New York, 1982, 329.
52. **Sakai, R., Tomiyama, K., Ishizaka, N., and Sato, N.,** Phenol metabolism in relation to disease resistance in potato tubers, *Ann. Phytopathol. Soc. Jp.,* 35, 207, 1967.
53. **Hammerschmidt, R.,** Rapid deposition of lignin in potato tuber tissue as a response to fungi non-pathogenic on potato, *Physiol. Plant Pathol.,* 24, 33, 1984.
54. **Chet, I., Hadar, Y., Elad, Y., Katan, J., and Henis, Y.,** Biological control of soilborne plant pathogens by *Trichoderma harzianum,* in *Soilborne Plant Pathogens,* Schippers, B. and Gams, W., Eds., Academic Press, London, 1979, 585.
55. **Kranz, J.,** Hyperparasitism of biotrophic fungi, in *Microbial Ecology of the Phylloplane,* Blakeman, J. P., Ed., Academic Press, London, 1981, 327.
56. **Tronsmo, A. and Raa, J.,** Antagonistic action of *Trichoderma pseudokoningii* against the apple pathogen *Botrytis cinerea, J. Phytopathol.,* 89, 216, 1977.
57. **Elad, Y., Chet, I., and Henis, Y.,** Degradation of plant pathogenic fungi by *Trichoderma harzianum, Can. J. Microbiol.,* 28, 719, 1982.
58. **Labudova, I. and Gogorova, L.,** Biological control of phytopathogenic fungi through lytic action of *Trichoderma* species, *FEMS Microbiol. Lett.,* 52, 193, 1988.
59. **Wisniewski, M., Biles, C., Droby, S., McLaughlin, R., Wilson, C., and Chalutz, E.,** Mode of action of the postharvest biocontrol yeast, *Pichia guilliermondii.* I. Characterization of attachment to *Botrytis cinerea, Physiol. Mol. Plant Pathol.,* 39, 245, 1991.
60. **Droby, S., Robin, D., Chalutz, E., and Chet, I.,** Possible role of glucanase and extracellular polymers in the mode of action of yeast antagonists of postharvest diseases, *Phytoparasitica,* 2, 167, 1993.
61. **Notario, V.,** β-glucanase from *Candida albicans*: purification, characterization and the nature of their attachment to cell wall components, *J. Gen. Microbiol.,* 128, 747, 1982.

Chapter 7

Enhancement of Biocontrol Agents for Postharvest Diseases and Their Integration with other Control Strategies

P. Lawrence Pusey

CONTENTS

I. INTRODUCTION

The potential of biological control in the postharvest environment has been discussed in several reviews.[1-6] Attempts to control foliar plant diseases in the field by the application of microbial antagonists have often failed.[8-10] Survival and growth of antagonists in field environments is probably limited due to water or nutrient deficiencies, harmful radiation, fluctuating climatic conditions, indigenous microorganisms being more competitive, or other factors. The primary advantage of postharvest biocontrol is the ability to control environmental conditions that determine the microecology on plant commodity surfaces. Because of this advantage, greater opportunities also exist after harvest to enhance the activity of biocontrol agents.

Research on biocontrol in the phyllosphere has evolved to the point that many workers believe further advances will come only when the leaf environment and epiphytic microbial communities are better understood.[8,10-11] Although the microecology on stored commodities is less complex than that on leaves in field environments, improvements in postharvest biocontrol are more likely to be made when we know how antagonists interact with the plant host, the environment, and other microorganisms. Also, an understanding of the mechanisms responsible for biocontrol is essential for correctly interpreting reasons for success or failure, and should be a prerequisite to its improvement.

Despite the demonstrated successes with microbial antagonists tested against various diseases of harvested crops,[1-2,4-7] even the maximum results attainable with these biocontrol agents may not match the efficacy of conventional fungicides that have been a mainstay in many fruit and vegetable industries. It will be particularly difficult to achieve a level of control equal to that with fungicides having systemic action. Biocontrol agents may provide protection on the surface or in fresh wounds of commodities, but they are unlikely to affect latent or incipient infections. Disease management that approaches the standards established with synthetic fungicides may be accomplished only through an integrated approach. In systems that include biocontrol agents, it will be important to also utilize to the greatest extent feasible conventional physical methods of reducing disease development (e.g., sanitation, handling practices that minimize injury to the commodity, and refrigeration). In additon, it might be possible to develop novel control strategies that will compliment biocontrol.

II. ENHANCING THE ACTIVITY OF MICROBIAL ANTAGONISTS

A. INDIGENOUS ANTAGONISTS

The use of microbial antagonists in postharvest disease control can be approached in two ways: (1) management of the indigenous microflora on the surfaces of produce and (2) artificial introduction of antagonists. The microecology of fruit and vegetable surfaces has not been studied extensively, and the possible manipulation of beneficial microorganisms present within the natural microflora on harvested crops is virtually an unexplored area.[4,12]

The premise that biological control occurs naturally on aerial surfaces of plants is supported by work with epiphytic populations on leaf surfaces[8-11,13-14] and other evidence that is largely circumstantial. Such other evidence has been provided by fungicides that affect microorganisms other than the pathogen at which they are directed. In many cases, diseases other than those being targeted appear to have increased because of a negative effect on resident antagonists that previously held a pathogen in check. Examples of such iatrogenic diseases include postharvest rots caused by increases in: *Rhizopus* spp. on strawberry after the use of benomyl to control *Botrytis cinerea*;[15-16] *Alternaria citri* on citrus after the use of benomyl to control *Penicillium* spp.[17] and *Colletotrichum gloeosporioides*;[18] and *Phytophthora syringae*[19] and *Alternaria tenuis*[20] on pome fruit after the use of benzimidazols to control *Gloeosporium* spp. and *Penicillium expansum.*

Another indication of natural biocontrol on the surfaces of harvested crops is the fact that washed commodities frequently develop more rot than unwashed commodities. Chalutz[21] found that washed and dried citrus fruit rotted more rapidly in storage than did unwashed fruit. They also discovered that bacteria and yeasts predominated when undiluted washings from citrus fruit were plated out. Only after the washings were diluted did fungal rot pathogens appear. It was suggested that epiphytic microbial populations on citrus naturally suppress rot pathogens.

In considering the potential for manipulating indigenous antagonists on produce, we can draw from studies dealing with leaf surfaces. Manipulation of chemical or nutritional variables on the phylloplane has been shown to alter populations of resident microorganisms. Morris and Rouse[22] determined the differential ability of epiphytic bacteria from snap bean leaves to utilize single carbon and nitrogen sources. By applying selective nutrients to foliage in field plots, they modified the composition of the bacterial community, altered the population size of fluorescent pseudomonads, and in some cases reduced disease caused by *Pseudomonas syringae*. In the postharvest environment, it further may be possible to favor certain antagonists by altering physical factors such as temperature, humidity, and atmospheric gases.

Conceptually, enhancement of beneficial epiphytic populations may also be achieved through genetic manipulation of the plant. Support for this approach includes evidence that the plant genotype has a role in survival and multiplication of pathogens. For example, workers have reported large differences in populations of virulent *Pseudomonas syringae*[23] or *Xanthomonas campestris* pv. *phaseoli*[24] on resistant and susceptible bean lines in the field. Marshall[25] reported that the genotype of oats was important in supporting populations of ice nucleation active bacteria. Bird[26] attributed the success of his multiple adversity resistance (MAR) cultivars of cotton to a combination of wide-adaptation, genetic resistance, and plant-associated microorganisms giving protection against disease and insect pests. Microbial populations isolated from surfaces of MAR cultivars contained more antagonists than those from susceptible varieties. Phenotypic characteristics that possibly affect the microflora include surface topography and exudate chemistry. It is conceivable then that plants could be manipulated through conventional genetic or recombinant deoxyribonucleic acid (DNA) methods to improve natural biological control on harvested crops. While attractive in theory, this approach is viewed at best as a long-range possibility.

Certainly, there is potential for enhancing indigenous antagonists on the surfaces of harvested crops. However, biological control with antagonists has so far been demonstrated mainly through the artificial introduction of large numbers of known antagonists. With the latter approach, the antagonists are directly manipulated, and thus there are more avenues for improvement.

B. INTRODUCED ANTAGONISTS
1. Production, Formulation, and Delivery

The main concern relating to production, whether by liquid or solid/semisolid surface fermentations, is that microbial inoculants are viable as colonizers and antagonists on plant surfaces.[27] Selective pressures that exist in laboratory cultures or in commercial production plants could result in the loss of characteristics important to the success of biocontrol. For example, structural features of bacteria that function in nature as attachment devices, permeability barriers, ion exchange resins, or protection against osmotic stress may be lost under conditions that prevail in laboratory cultures.[28-29] It has been found that the bacterial glycocalyx is lost when bacteria are grown in many laboratory media,[30] but this can sometimes be prevented by altering cultural conditions.[31] The ability of microbes to produce certain metabolites may also be affected during the fermentation process. Lindow[32] showed that strains of bacteria antagonistic to *Pseudomonas syringae* on rich culture media did not produce inhibitory compounds when grown on a medium designed to simulate the nutrient conditions of leaf surfaces.

Methods of production that maintain or increase the viability of antagonists may result in more effective biocontrol. However, there has been insufficient research in this area. Concepts dealing with the mass production of microorganisms have been modeled after pharmaceutical products rather than microbial agents for reintroduction to plant surfaces. Therefore, biocontrol organisms present a unique challenge to industry because of the need for mass quantities of "competent" inoculum that is compatible with existing storage and application technology.[8,33] Production efficiency and cost must of course be considered before applying new technology on a commercial scale.

Maintaining viability is also fundamental in formulating antagonists as commercial products. Biocontrol agents must have a storage life of at least 6 months and preferably 1 to 2 years.[34-39] Perhaps recent advances in encapsulation technology using cross-linked matrix organic polymers[34,37,40-42] will increase the shelf life of products. Some types of additives that serve an important function in harsh field environments (e.g., to protect against desiccation or ultraviolet radiation) may not be necessary in the postharvest environment. However, utilization of adjuvants such as nutrients, pH buffers, or other substances to stimulate or enhance microbial activity on stored commodities appears promising. In addition, formulation technology developed to improve the dispersal and adherence of agents to plant surfaces in the field may have application in postharvest systems.

Formulation and delivery should be viewed together as one system. In deciding what application method to employ, a logical starting point is to consider delivery systems already in place for applying synthetic fungicides to harvested crops. Fungicide treatments are made via dip tanks, hydrocooling water, over-line spray or mist, and brushes.

In the case of many tree fruits, it is now a common practice to apply fungicides mixed with coating wax. The wax also serves to reduce moisture loss and improve commodity appearance. There has been recent interest in natural edible coatings that extend shelf life;[43-46] thus, other materials could eventually replace the wax coatings currently being used. There are two basic kinds of waxing systems: (1) those involving water-based wax applied by over-line spray and (2) others consisting of oil-based wax (e.g., paraffin and mineral oil) that is drip-fed onto brushes that come into contact with the commodity. The first method is widely used, but is generally less efficient because of wasted runoff. Equipment is sometimes designed to recycle the material, but this often leads to an accumulation of debris and pathogen inoculum that is reintroduced onto produce. With the second system, essentially the only material wasted is what remains on the equipment. (Researchers tend to avoid work with the latter system because it is next to impossible to remove wax and experimental agents between treatments.) Both types of wax have been commercially preformulated with fungicides approved for the commodity for which they are designed. Packers of produce have the option of using the preformulated material or mixing fungicide with wax themselves. Similarly, biocontrol agents could be incorporated into wax either before or after these materials are in the hands of packinghouse workers. Antagonistic microorganisms might be better preserved in oil-based waxes, since mineral oil has been used successfully for many years in the storage of fungal and bacterial cultures.[47]

Waxes and other fruit coatings could positively or negatively affect the activity of biocontrol agents. Wilson and Wisniewski[47a] found that yeast antagonists performed much better when applied prior to coating fruit with wax, as opposed to treating once with a yeast-wax mixture. Pusey et al.[48,49] found that waxes had no influence on the effect of *Bacillus subtilis* against brown rot of peaches, whether the wax and the bacterium were applied separately or as a mixture.

For commodities and postharvest systems that do not include wax or for those systems in which it is best to apply the biocontrol agent apart from wax, other delivery methods must be developed. Use of conventional spray equipment has certain disadvantages, as mentioned above. A rotary disk atomization system recently developed by Wilson and Wisniewski[47a] dispenses a concentrated preparation of the antagonist in metered amounts, with minimal waste. All systems developed for delivering biocontrol agents will have to be evaluated with respect to their effect on antagonist viability and on the placement of these organisms for optimal results.

2. Manipulation of Microenvironment

The potential for enhancing postharvest biocontrol by manipulating the environment is immense. Temperature, humidity, and gas composition can easily be controlled in storage chambers. Of course, conditions currently imposed after harvest are based on established optima for maximizing physiological shelf life of the commodity and suppressing disease agents. The benefits of any change that favors antagonists will have to be weighed against the possible negative effects as related to other factors. Perhaps the greatest opportunities to enhance biocontrol via the environment lie with the nutritional or chemical milieu, which could be altered to give antagonists a selective advantage or increase their activity on commodity surfaces.

As already discussed, the addition of nutrients to foliage or other plant surfaces can dramatically alter epiphytic populations. Nutritional amendments have also been shown to enhance antagonists introduced onto leaf surfaces. For example, Fokkema et al.[50] found that yeasts (*Cryptococcus* and *Sporobolomyces*), when sprayed onto wheat leaves in combination with 2% sucrose and 0.1% yeast extract, increased in number compared to controls involving yeasts applied in water. The pathogens, *Septoria* and *Cochliobolus*, were reduced initially; and disease development was slowed in the early stages.

Gullino et al.[51] found that the suppression of *Botrytis cinerea* by *Trichoderma* spp. on detached grape clusters was greater when malt extract was added to the antagonist inoculum. In efforts to improve postharvest biocontrol of blue mold of apples, Janisiewicz et al.[52] evaluated 36 carbohydrates and 23 nitrogenous compounds for their effect on germination and growth of the pathogen, *Penicillium expansum*, and for stimulation of growth of an antagonistic strain of *Pseudomonas syringae*. Compounds that strongly stimulated growth of the antagonist but had little or no stimulatory effect on the pathogen were selected for further testing to evaluate their effect on biocontrol. L-Asparagine and L-proline increased the population of the antagonist on fruit by more than one order of magnitude and enhanced biocontrol of blue mold.

Nutrient composition not only can affect the population density and competitiveness of antagonists, but also it can influence the production of metabolites that are crucial in many postharvest biocontrol systems. Such systems include those involving antibiotics[53-54] or cell wall degrading enzymes.[55] The synthesis of antibiotics by microorganisms is under nitrogen regulation and subject to catabolite repression.[56-58] The form and concentration of nitrogen and carbon may be important to the synthesis and excretion of compounds that are a key to successful biocontrol.

McLaughlin et al.[59] discovered that certain salts enhance the activity of yeast antagonists against *Botrytis cinerea* and *Penicillium expansum* on harvested apples. Calcium chloride at a 2% concentration was the most effective salt. The phenomenon is unrelated to osmotic potential and not yet understood. Enhancement of postharvest biocontrol with calcium chloride has now been demonstrated with several different yeast genera and with various pathogen and commodity systems.[60,61]

Recent work indicates that defined organic amendments can improve biocontrol on aerial surfaces of plants. Kokalis-Burelle et al.[62] showed that chitin applied to peanut leaves effectively sustained high populations of a chitinolytic strain of *Bacillus cereus*. The treatment resulted in a significant reduction in early leaf spot caused by *Cercospora arachidicola*. Davis et al.[63] successfully used hydrolyzed cellulose and a vegetable oil-based spreader-sticker to improve the adhesion and activity of the antagonist *Chaetomium globosum* for control of apple foliar diseases. Similarly, Tronsmo[64] improved the control of apple diseases with carboxymethyl cellulose applied with the antagonist *Trichoderma harzianum*.

3. Mixtures of Antagonists

According to Baker and Cook,[65] "there is a better chance of attaining successful biological control with a mixture of several antagonists than with a single one." It is a general principle that complex associations in nature are more stable. The application of this axiom must of course involve antagonists that are complementary and not competitive.

Utilization of more than one antagonist could be particularly advantageous when the combination of organisms represents a diversity of biocontrol mechanisms. This approach may reduce the chances that pathogens will, as the result of selection pressure, overcome the inhibition caused by biocontrol agents.

It must also be taken into account that one antagonist may not effectively control all important pathogens of a given commodity. Even single chemical fungicides are often inadequate. For example, at least two different fungicides have conventionally been used in combination for controlling postharvest diseases of stone fruit. This is because fungicides effective against brown rot of stone fruit offer little protection against *Rhizopus* rot, another important postharvest disease of these crops. The one fungicide that controls *Rhizopus* rot (i.e., dicloran) does not affect brown rot. Janisiewicz[66] reported that a mixture of two antagonists, *Acremonium breve* and *Pseudomonas* sp., gave total protection against postharvest apple diseases caused by *Botrytis cinerea* and *Penicillium expansum*.

4. Genetic Enhancement

Although it is conceivable that genetic manipulation of the host plant might improve biocontrol, manipulation of the antagonist may be a more feasible and realistic approach. It is well known that genetic variability is high among strains of antagonistic microorganisms.[67-69] Genes that impart high competency or potency to antagonists may be utilized through selection or recombinant DNA technology. Desirable gene-mediated characteristics of antagonists could include the addition or an elevated degree of the following: (1) utilization of compounds of host origin, (2) utilization of applied compounds that favor antagonists, (3) tolerance to pesticides, (4) colonization of host surfaces under storage conditions, and (5) synthesis of compounds that contribute to antagonism (e.g., antibiotics, enzymes, siderophores, or receptor proteins for siderophore uptake).

Abd-El Moity et al.,[70] Papavizas and Lewis,[71] and Papavizas et al.[72] used ultraviolet light induction followed by selection to produce isolates of *Trichoderma* species with tolerance to fungicides and enhanced biocontrol capabilities against soilborne diseases. Tronsmo[73] improved the biocontrol of *Botrytis cinerea* on apples by selecting *Trichoderma* isolates capable of growth at the relatively low temperatures that exist during the flowering period when infection is initiated. Similar approaches could be taken in exploiting the genetic potential of microbial populations that are antagonistic toward postharvest pathogens.

Recombinant methods enable genes to be introduced with considerable precision from outside the normal gene pool of the population. Theoretically, characteristics from more than one population could be combined in the same organism to increase antagonist capability. For instance, an organism that is a good colonizer could be the recipient of genes coding for the synthesis of specific antibiotics or enzymes. It may also be possible to improve the performance of antagonists by altering the regulation of metabolite synthesis. Gutterson[74] and Gutterson and co-workers[75] made transcriptional fusions to enhance oomycin A synthesis in *Pseudomonas fluorescens* for control of *Pythium ultimum.* Such genetic manipulations will likely be limited to bacterial antagonists in the near future, since the genetics of fungi are much less understood.

III. INTEGRATING BIOCONTROL WITH OTHER CONTROL STRATEGIES

The production of fresh fruits and vegetables and their movement from the farm to the consumer are a complex chain of events. Adequate disease control is dependent on a number of measures employed along the length of this chain. In the absence of synthetic fungicides, an integrated approach to control will be especially important.

A. COMPATIBILITY WITH CONVENTIONAL PRACTICES

The compatibility of antagonists with current commercial practices and conditions should be considered in developing biocontrol agents for postharvest use. It is desirable that antagonists be unaffected by chemical residues (from pre- or postharvest applications) and widely used wax coatings; and that they can survive temperature, humidity, and atmospheric conditions during storage. As mentioned earlier, any alteration in environment to favor the antagonist must be viewed with consideration to what effects the change will have on the host and pathogen. *Bacillus subtilis* (strain B-3), an effective antagonist against brown rot of stone fruit,[76] was found to be compatible with commercial waxes, dicloran (used for control of *Rhizopus* rot), and cold storage temperatures.[48] Compatibility between a microbial antagonist and a synthetic fungicide allows the option of using the antagonist in combination with reduced levels of the fungicide. In pilot tests involving the application of the yeast antagonist *Pichia guilliermondii* (strain US-7) to citrus fruit,[77] yeast combined with the fungicide thiabendazole (TBZ) at one tenth the recommended rate reduced decay to a level equal to that of the current commercial treatment of TBZ at the full rate.

B. NOVEL METHODS THAT MAY COMPLIMENT BIOCONTROL

A number of strategies for reducing postharvest disease losses are currently being investigated as alternatives to conventional fungicides. Some of these approaches are relatively new while others have received renewed interest because of sociopolitical pressures resulting in the loss or possible loss of fungicides from postharvest use. Methods being considered alone or as part of combinations include: heat treatments,[78,79] induced resistance with ultraviolet light,[80-82] calcium infiltration,[83-86] treatments with generally regarded as safe (GRAS) compounds commonly used as food additives,[87,88] and natural products derived from plants.[89,90] These and other methods might be used with biocontrol agents to produce additive or synergistic effects against disease.

Although microbial antagonists may be effective when present before or soon after the arrival of pathogens on commodity surfaces, they are unlikely to control latent or incipient infections.[91] This weakness of biocontrol could be compensated for by other control methods within an integrated sytem. Heat treatments do not impart residual action (unless host resistance is induced) as is expected with antagonists; however, heat applied to some commodities may compliment biocontrol by killing organisms that have penetrated the epidermis or outer cell layers. Sterilizing the commodity surface with heat prior to the addition of the biocontrol agent might also give the antagonist an advantage since competition from other microorganisms would be eliminated. This is analogous to pasteurizing soil prior to adding an antagonist, a strategy that reportedly enhances the biocontrol of some soilborne diseases.[92-95]

The success rate of biocontrol may also be improved through integration with measures that increase host resistance (e.g., classical breeding and selection, induced resistance, or genetic engineering). Such resistance would reduce the incidence and activation of latent infections of inner tissues that are "out of reach" for the antagonist.

Antifungal substances (GRAS compounds or natural plant products) that selectively inhibit pathogens could also be used in combination with biocontrol agents. Pathogens might be weakened and become more vulnerable to antagonist activity. The natural edible coatings that might be used on produce in the future[43-46] could be "designed" to discourage pathogens through inhibitory action and/or encourage biocontrol agents by providing a nutritional and physical microenvironment conducive to their activity.

The possibilities and opportunities for integrated disease management involving biocontrol agents appear unlimited.

IV. CONCLUDING REMARKS

We have witnessed a growing public concern in recent years regarding the safety of pesticides used to protect agricultural crops. The pressure to eliminate synthetic fungicides from postharvest use has been particularly great. Biological control appears to be a viable alternative. Because conditions and factors affecting microbial colonization on commodity surfaces can be controlled to a high degree after harvest, the opportunities to enhance the activity of antagonists are exceptionally good. At the same time, it is doubtful that biocontrol agents by themselves will ever provide the efficacy or consistency associated with conventional fungicides. However, the probability is increasing that they will eventually constitute a significant part of integrated systems that can provide adequate control of postharvest diseases.

REFERENCES

1. **Jeffries, P. and Jeger, M. J.,** The biological control of postharvest diseases of fruit, *Postharvest News Inf.,* 1, 365, 1990.
2. **Pusey, P. L., Wilson, C. L., and Wisniewski, M. E.,** Management of postharvest diseases of fruits and vegetables: strategies to replace vanishing fungicides, in *Pesticide-Plant Interactions in Crop Production: Beneficial and Deleterious Effects,* Altman, J., Ed., CRC Press, Boca Raton, FL, 1993, chap. 26.
3. **Wilson, C. L. and Pusey, P. L.,** Potential for biological control of postharvest plant diseases, *Plant Dis.,* 69, 375, 1985.
4. **Wilson, C. L. and Wisniewski, M. E.,** Biological control of postharvest diseases of fruits and vegetables: an emerging technology, *Annu. Rev. Phytopathol.,* 27, 425, 1989.
5. **Wilson, C. L., Wisniewski, M. E., Biles, C. L., McLaughlin, R., Chalutz, E., and Droby, S.,** Biological control of post-harvest diseases of fruits and vegetables: alternatives to synthetic fungicides, *Crop Prot.,* 10, 172, 1991.
6. **Wisniewski, M. E. and Wilson, C. L.,** Biological control of postharvest diseases of fruits and vegetables: recent advances, *HortScience,* 27, 94, 1992.
7. **Janisiewicz, W.,** Biological control of postharvest fruit diseases, in *Handbook of Applied Mycology,* Vol. 1, Arora, D. K., Rai, B., Muderji, K. G., and Knudsen, G. R., Eds., Marcel Dekker, New York, chap. 12.
8. **Andrews, J. H.,** Biological control in the phyllosphere: realistic goal or false hope?, *Can. J. Plant Pathol.,* 12, 300, 1990.
9. **Andrews, J. H.,** Biological control in the phyllosphere, *Annu. Rev. Phytopathol.,* 30, 603, 1992.
10. **Knudsen, G. R. and Spurr, H. W., Jr.,** Management of bacterial populations for foliar disease biocontrol, in *Biocontrol of Plant Diseases,* Vol. 1, Mukerji, K. G. and Garg, K. L., Eds., CRC Press, Boca Raton, FL, 1988, chap. 6.
11. **Spurr, H. W., Jr. and Knudsen, G. R.,** Biological control of leaf diseases with bacteria, in *Biological Control on the Phylloplane,* Windels, C. E. and Lindow, S. E., Eds., American Phytopathological Society, St. Paul, MN, 1985, 45.
12. **Wilson, C. L.,** Managing the microflora of harvested fruits and vegetables to enhance resistance, *Phytopathology,* 79, 1387, 1989.
13. **Upper, C. D.,** Manipulation of microbial communities in the phyllosphere, in *Microbial Ecology of Leaves,* Andrews, J. H. and Hirano, S. S., Springer-Verlag, New York, 1992, chap. 23.
14. **Fokkema, N. J.,** The phyllosphere as an ecologically neglected milieu: a plant pathologist's point of view, in *Microbial Ecology of Leaves,* Andrews, J. H. and Hirano, S. S., Springer-Verlag, New York, 1992, chap. 1.
15. **Dennis, C.,** Effect of pre-harvest fungicides on the spoilage of soft fruit after harvest, *Ann. Appl. Biol.,* 81, 227, 1975.
16. **Jordan, V. W. L.,** The effects of prophylactic spray programs on the control of pre- and post-harvest diseases of strawberry, *Plant Pathol.,* 22, 67, 1973.
17. **Singh, V.,** Control of *Alternaria* rot in citrus fruit, *Australas, Plant Pathol.,* 9, 12, 1980.
18. **Laville, E.,** Evolution des pourritures d'entreposage des argrumes avec l'utilisation de nouveaux fongicides de traitement apres recolte, *Fruits,* 26, 33, 1971.
19. **Upstone, M.,** Evaluation of chemicals for control of *Phytophtora* fruit rot in stored apples, *Proc. Br. Crop Prot. Conf. Pest Dis.,* 1, 203, 1977.
20. **Spalding, D. H.,** Postharvest use of benomyl and thiabendazole to control blue-mold rot development in pears, *Plant Dis. Rep.,* 54, 655, 1970.
21. **Chalutz, E. and Wilson, C. L.,** Postharvest biocontrol of green and blue mold and sour rot of citrus fruit by *Debaryomyces hansenii, Plant Dis.,* 74, 134, 1990.

22. **Morris, C. E. and Rouse, D. I.,** Role of nutrients in regulating epiphytic bacterial populations, in *Biological Control on the Phylloplane,* Windels, C. E. and Lindow, S. E., Eds., American Phytopathological Society, St. Paul, MN, 1985, 63.

23. **Daub, M. E. and Hagedorn, D. J.,** Epiphytic populations of *Pseudomonas syringae* on susceptible and resistant bean lines, *Phytopathology,* 71, 547, 1981

24. **Cafati, C. R. and Saettler, A. W.,** Effect of host on multiplication and distribution of bean common blight bacteria, *Phytopathology,* 70, 675, 1980.

25. **Marshall, D.,** A relationship between ice-nucleation-active bacteria, freeze damage and genotype in oats, *Phytopathology,* 78, 952, 1988.

26. **Bird, L. S.,** The MAR (multi-adversity resistance) system for genetic improvement of cotton, *Plant Dis.,* 66, 172, 1982.

27. **Lethbridge, G.,** An industrial view of microbial inoculants for crop plants, in *Microbial Inoculation of Crop Plants,* Campbell, R. and Macdonald, R. M., Eds., IRL Press, Oxford, 1989, chap. 2.

28. **Nikaido, H. and Vaara, M.,** Molecular basis of bacterial outer membrane permeability, *Microbiol. Rev.,* 49, 1, 1985.

29. **Sleyter, U. B. and Messner, P.,** Crystalline surface layers on bacteria, *Annu. Rev. Microbiol.,* 37, 311, 1983.

30. **Costerton, J. W., Geesey, G. G., and Cheng, K. J.,** How bacteria stick, *Sci. Am.,* 238, 86, 1978.

31. **Govan, J. R. W.,** Mucoid strains of *Pseudomonas aeruginosa:* the influence of culture medium on the stability of mucus production, *J. Med. Microbiol.,* 8, 513, 1975.

32. **Lindow, S. E.,** Lack of correlation of *in vitro* antbiosis with antagonism of ice nucleation active bacteria on leaf surfaces by non-ice nucleation active bacteria, *Phytopathology,* 78, 444, 1988.

33. **Lisansky, S. G. and Hall, R. A.,** Fungal control of insects, in *The Filamentous Fungi,* Vol. 4, Fungal Technology, Smith, J. E., Berry, D. R., and Kristiansen, B., Eds., Edward Arnold, London, 1983, 327.

34. **Baker, C. A. and Henis, J. M. S.,** Commercial production and formulation of microbial biocontrol agents, in *New Directions in Biological Control: Alternatives for Suppressing Agricultural Pests and Diseases,* Baker, R. R. and Dunn, P. E., Eds., Alan R. Liss, New York, 1990, 333.

35. **Caesar, A. J. and Burr, T. J.,** Effect of conditioning, betaine, and sucrose on survival of rhizobacteria in powder formultations, *Appl. Environ. Microbiol.,* 1, 23, 1991.

36. **Carlton, B. C.,** Economic considerations in marketing and application of biocontrol agents, in *New Directions in Biological Control: Alternatives for Suppressing Agricultural Pests and Diseases,* Baker, R. R. and Dunn, P. E., Eds., Alan R. Liss, New York, 1990, 419.

37. **Connick, W. J., Jr., Lewis, J. A., and Quimby, P. C., Jr.,** Formulation of biocontrol agents for use in plant pathology, in *New Directions in Biological Control: Alternatives for Suppressing Agricultural Pests and Diseases,* Baker, R. R. and Dunn, P. E., Eds., Alan R. Liss, New York, 1990, 346.

38. **Macdonald, R. M.,** An overview of crop inoculation, in *Microbial Inoculation of Crop Plants,* Campbell, R. and Macdonald, R. M., Eds., IRL Press, Oxford, 1989, 1.

39. **McIntyre, J. L. and Press, L. S.,** Formulation, delivery systems and marketing of biocontrol agents and plant growth promoting rhizobacteria (PGPR), in *The Rhizosphere and Plant Growth,* Keister, D. L. and Cregan, P. B., Eds., Kluwer, Dordrecht, 1991, 289.

40. **Harman, G. R.,** Deployment tactics for biocontrol agents in plant pathology, in *New Directions in Biological Control: Alternatives for Suppressing Agricultural Pests and Diseases,* Baker, R. R. and Dunn, P. E., Eds., Alan R. Liss, New York, 1990, 779.

41. **Lewis, J. A.,** Formulation and delivery systems of biocontrol agents with emphasis on fungi, in *The Rhizosphere and Plant Growth,* Keister, D. L. and Cregan, P. B., Eds., Kluwer, Dordrecht, 1991, 279.

42. **Lewis, J. A. and Papavizas, G. C.,** Biocontrol of plant diseases: the approach for tomorrow, *Crop Prot.,* 10, 95, 1991.

43. **Dhalla, R. and Hanson, S. W.,** Effect of permeable coatings on the storage life of fruits. II. Pro-long treatment of mangoes (*Mangifera indica* L. cv. Julie), *Int. J. Food Sci. Technol.,* 23, 107, 1988.

44. **Kester, J. J. and Fennema, O. R.,** Edible films and coatings: a review, *Food Technol.,* 40, 47, 1986.

45. **Nisperos, M. O. and Baldwin, E. A.,** Effect of two types of edible films on tomato fruit ripening, *Proc. Fla. State Hortic. Soc.,* 101, 217, 1988.

46. **Nisperos-Carriedo, M. O., Shaw, P. E., and Baldwin, E. A.,** Changes in volatile flavor components of pineapple orange juice as influenced by the application of lipid and composite films, *J. Agric. Food Chem.,* 38, 1382, 1990.

47. **Dhingra, O. D. and Sinclair, J. B.,** *Basic Plant Pathology Methods,* CRC Press, Boca Raton, FL, 1985, chap. 3.

47a. **Wilson, C. L. and Wisniewski, M.,** personal communication.

48. **Pusey, P. L., Wilson, C. L., Hotchkiss, M. W., and Franklin, J. D.,** Compatibility of *Bacillus subtilis* for postharvest control of peach brown rot with commercial fruit waxes, dicloran, and cold-storage conditions, *Plant Dis.,* 70, 587, 1986.

49. **Pusey, P. L., Hotchkiss, M. W., Dulmage, H. T., Baumgardner, R. A., Zehr, E. I., Reilly, C. C., and Wilson, C. L.,** Pilot tests for commercial production and application of *Bacillus subtilis*(B-3) for postharvest control of peach brown rot, *Plant Dis.,* 72, 622, 1988.

50. **Fokkema, N. J., Den Houter, J. G., Kosterman, Y. J. C., and Nelis, A. L.,** Manipulation of yeasts of field-grain wheat leaves and their antagonistic effect on *Cochliobolis sativus* and *Septoria nodorum,Trans. Br. Mycol. Soc.,* 72, 19, 1979.

51. **Gullino, M. L., Aloi, C., and Garibaldi, A.,** Evaluation of the influence of different temperatures, relative humidities and nutritional supports on the antagonistic activity of *Trichoderma* spp. against grey mold of grape, in *Influence of Environmental Factors on the Control of Grape Pests, Diseases, and Weeds,* Proc. EC Expert's Group, Cavalloro, R., Ed., Instituto di Patologia Vegetale dell Universita, Torino, Italy, 1989, 231.

52. **Janisiewicz, W. J., Usall, J., and Bors, B.,** Nutritional enhancement of biocontrol of blue mold on apples, *Phytopathology,* 82, 1364, 1992.

53. **Pusey, P. L.,** Use of *Bacillus subtilis* and related organisms as biofungicides, *Pestic. Sci.,* 27, 133, 1989.

54. **Janisiewicz, W. J. and Roitman, J.,** Biological control of blue mold and gray mold on apple and pear with *Pseudomonas cepacia, Phytopathology,* 78, 1697, 1988.

55. **Wisniewski, M., Biles, C., Droby, S., McLaughlin, R., Wilson, C., and Chalutz, E.,** Mode of action of the postharvest biocontrol yeast, *Pichia guilliermondii.* I. Characterization of attachment to *Botrytis cinerea,Physiol. Mol. Plant Pathol.,* 40, 1992.

56. **Aharonowitz, Y.,** Nitrogen metabolite regulation of antibiotic biosynthesis, *Annu. Rev. Microbiol.,* 34, 209, 1980.

57. **Malik, V. S.,** Genetics and biochemistry of secondary metabolism, *Adv. Appl. Microbiol.,* 28, 27, 1982.

58. **Nagato, N., Okumura, Y., Okamoto, R., and Ishikura, T.,** Carbon catabolite regulation of neoviridogrisein production by glucose, *Agric. Biol. Chem.,* 48, 3041, 1984.

59. McLaughlin, R. J., Wisniewski, M. E., Wilson, C. L., and Chalutz, E., Effect of inoculum concentration and salt solutions on biological control of postharvest diseases of apple with *Candida* sp., *Phytopathology*, 80, 456, 1990.

60. Gullino, M. L., Aloi, C., Palitto, M., and Benzi, D., Attempts at biological control of postharvest diseases of apple, *Phytoparasitica*, 19, 258, 1991.

61. McLaughlin, R. J., Wilson, C. L., Droby, S., Ben-Arie, R., and Chalutz, E., Biological control of postharvest diseases of grape, peach, and apple with the yeasts *Kloeckera apiculata* and *Candida guilliermondii*, *Plant Dis.*, 76, 470, 1992.

62. Kokalis-Burelle, N., Backman, P. A., Rodriguez, R., and Ploper, D. L., Chitin as a foliar amendment to modify microbial ecology and control disease, *Phytopathology*, 81, 1152, 1991.

63. Davis, R. F., Backman, P. A., Rodriguez-Kabana, R., and Kokalis-Burelle, N., Biological control of apple fruit diseases by *Chaetomium globosum* formulation containing cellulose, *Biol. Control*, 2, 118, 1992.

64. Tronsmo, A., Biological and integrated controls of *Botrytis cinerea* on apple with *Trichoderma harzianum*, *Biol. Control*, 1, 59, 1991.

65. Baker, K. and Cook, R. J., *Biological Control of Plant Pathogens*, The American Phytopathological Society, St. Paul, MN, 1982, chap. 5.

66. Janisiewicz, W. J., Biocontrol of postharvest diseases of apples with antagonist mixtures, *Phytopathology*, 78, 194, 1988.

67. Cullen, D. and Andrews, J. H., Evidence for the role of antibiosis in the antagonism of *Chaetomium globosum* to the apple scab pathogen, *Venturia inaequalis*, *Can. J. Bot.*, 62, 1819, 1984.

68. Dubos, B., Fungal antagonism in aerial agrobiocenoses, in *Innovative Approaches to Plant Disease Control*, Chet, I., Ed., John Wiley & Sons, New York, 1987, 107.

69. Lindow, S. E., Competitive exclusion of epiphytic bacteria by ice *Pseudomonas syringae* mutants, *Appl. Environ. Microbiol.*, 53, 2520, 1987.

70. Abd-El Moity, T. H., Papavizas, G. C., and Shatla, M. N., Induction of new isolates of *Trichoderma harzianum* tolerant to fungicides and their experimental use for control of white rot of onion, *Phytopathology*, 72, 396, 1982.

71. Papavizas, G. C. and Lewis, J. A., Physiological and biocontrol characteristics of stable mutants of *Trichoderma viride* resistant to MBC fungicides, *Phytopathology*, 73, 407, 1983.

72. Papavizas, G. C., Lewis, J. A., and Abd-El Moity, T. H., Evaluation of new biotypes of *Trichoderma harzianum* for tolerance to benomyl and enhanced biocontrol capabilities, *Phytopathology*, 72, 126, 1982.

73. Tronsmo, A. and Ystaas, J., Biological control of *Botrytis cinerea* on apple, *Plant Dis.*, 64, 1009, 1980.

74. Gutterson, N., Microbial fungicides: recent approaches to elucidating mechanisms, *Crit. Rev. Biotechnol.*, 10, 69, 1990.

75. Gutterson, N., Howie, W., and Suslow, T., Enhancing efficiencies of biocontrol agents by use of biotechnology, in *New Directions in Biological Control: Alternatives for Suppressing Agricultural Pests and Diseases*, Baker, R. R. and Dunn, P. E., Eds., Alan R. Liss, New York, 1990, 749.

76. Pusey, P. L. and Wilson, C. L., Postharvest biological control of stone fruit brown rot by *Bacillus subtilis*, *Plant Dis.*, 68, 753, 1984.

77. Droby, S., Chalutz, E., Hofstein, R., Wilson, C. L., Wisniewski, M., Fridlender, B., Cohen, L., Weiss, B., and Daus, A., Pilot testing of *Pichia guilliermondii*: a biocontrol agent of postharvest diseases of citrus fruit, *Plant Dis.*, in press.

78. Barkai-Golan, R. and Douglas, J. P., Postharvest heat treatment of fresh fruits and vegetables for decay control, *Plant Dis.*, 75, 1085, 1991.

79. **Couey, H. M.,** Heat treatment for control of postharvest diseases and insect pests of fruits, *HortScience,* 24, 198, 1989.

80. **Lu, J. Y., Stevens, C., Khan, V. A., Kabwe, M., and Wilson, C. L.,** The effect of ultraviolet irradiation on shelf-life and ripening of peaches and apples, *J. Food Qual.,* 14, 299, 1991.

81. **Lu, J. Y., Stevens, C., Yakabu, P., Loretan, P. A., and Eakin, D.,** Gamma, electron beam and ultraviolet radiation on control of storage rots and quality of Walla Walla onions, *J. Food Process. Preserv.,* 12, 53, 1987.

82. **Stevens, C., Khan, V. A., Tang, A. Y., and Lu, J.,** The effect of ultraviolet radiation on mold rots and nutrients of stored sweet potatoes, *J. Food Prot.,* 53, 223, 1990.

83. **Conway, W. S., Greene, G. M., and Hickey, K. D.,** Effects of preharvest and postharvest calcium treatments of peaches on decay caused by *Monilinia fructicola, Plant Dis.,* 71, 1084, 1987.

84. **Conway, W. S. and Sams, C. E.,** Calcium infiltration of Golden Delicious apples and its effect on decay, *Phytopathology,* 73, 1068, 1983.

85. **Conway, W. S., Sams, C. E., Abbott, J. A., and Bruton, B. D.,** Postharvest calcium treatment of apple fruit to provide broad-spectrum protection against postharvest pathogens, *Plant Dis.,* 75, 620, 1991.

86. **McGuire, R. G. and Kelman, A.,** Reduced severity of *Erwinia* soft rot in potato tubers with increased calcium content, *Phytopathology,* 74, 1250, 1984.

87. **Bigelis, R.,** Fungal metabolites in food processing, in *Handbook of Applied Mycology,* Vol. 3, Arora, D. K., Mukerji, K. G., and Marth, E. H., Eds., Marcel Dekker, New York, 1991, chap. 13.

88. **Liewen, M. B.,** Antifungal food additives, in *Handbook of Applied Mycology,* Vol. 3, Arora, D. K., Mukerji, K. G., and Marth, E. H., Eds., Marcel Dekker, New York, 1991, chap. 16.

89. **Ghaouth, A. E., Arul, J., Grenier, J., and Asselin, A.,** Antifungal activity of chitosan on two postharvest pathogens of strawberry fruits, *Phytopathology,* 82, 398, 1992.

90. **Sholberg, P. L. and Shimizu, B. N.,** Use of the natural plant products, hinokitiol, to extend shelf-life of peaches, *Can. Inst. Sci. Technol. J.,* 24, 273, 1991.

91. **Eckert, J. W.,** Role of chemical fungicides and biological agents in postharvest disease control, in *Proc. Workshop Biol. Control Postharvest Dis. Fruits Vegetables,* Wilson, C. L. and Chalutz, E., Eds., Agricultural Research Service, U. S. Department of Agriculture, ARS-92, 1991, 14.

92. **Chet, I., Elad, Y., Kalfon, A., Hadar, Y., and Katan, J.,** Integrated control of soilborne and bulbborne pathogens in iris, *Phytoparasitica,* 10, 229, 1982.

93. **Elad, Y., Katan, J., and Chet, I.,** Physical, biological, and chemical control integrated for soilborne diseases in potatoes, *Phytopathology,* 70, 418, 1980.

94. **Katan, J.,** Solar pasteurization of soils for disease control: status and prospects, *Plant Dis.,* 64, 450, 1980.

95. **Strashnow, Y., Elad, Y., Sivan, A., and Chet, I.,** Integrated control of *Rhizoctonia solani* by methyl bromide and *Trichoderma harzianum, Plant Pathol.,* 34, 146, 1985.

Chapter 8

Large-Scale Production and Pilot Testing Biological Control Agents for Postharvest Diseases

Raphael Hofstein, Bertold Fridlender, Edo Chalutz, and Samir Droby

CONTENTS

I. INTRODUCTION

A bridge between theory and practice can be constructed when, and only when, a microbial pesticide performs its function reproducibly in a commercial environment. The vehicle that is always used for the transaction, from an event taking place in the petri dish to a profound production process, includes all that is involved in the tedious process of scaling up. The concept of scaling up of biopesticide production stems from the general theme of the biological sciences. It all starts with basic research that focuses on the fundamental phenomena of biological processes. These are reflected in the biotechnological and industrial events in terms of a functional microorganism(s) which is assigned to act as the executing machinery of the process.

The area of biological control of postharvest disease is still in its infancy, and there is very little information on which to establish concepts and guidelines for industrialization. The initial step in the development of a commercial agent for biological control is fulfillment by the potential antagonist of certain criteria on the basis of laboratory studies.

Development of a biofungicide for the control of postharvest disease is one approach that is relatively easy to analyze and adjust an effective antagonist in the isolated environment of a research laboratory to the "real world" of packing houses. The investment of efforts and financial resources in such a development is justified by two major considerations: the first and predominant one is that most chemical fungicides were withdrawn from the market for ecological and medical reasons; the second is that due to pesticide resistance, most of the commonly used chemicals became less effective and in turn were applied in higher rates. This created a problem of cost-effectiveness as well as a new dimension of ecological hazards.

Moreover, the postharvest environment appears to present a better milieu for biological control than do field conditions. In the postharvest environment it is often possible to control temperature and humidity. In addition, the postharvest environment is an artificial "ecological island" separated from the buffering effect of natural microbial ecosystems.

0-8493-4567-7/94/$0.00+$.50
© 1994 by CRC Press Inc.

Such conditions favor the use of introduced antagonists for biological control. These advantages were extensively discussed by Wilson et al.[1]

The postharvest processing industry presents a case for alternative methods of pest control, and the proposal to use biological control of the diseases has been extensively discussed as a promising option by Wilson and Wisniewski.[2] Another angle of the situation was viewed by Spurr et al.[3] who claimed that in order to "...achieve biological control it is necessary to do biological research without promising biological control!" Industry does not have that latitude for too long a period of time. They must produce a commercially successful product within a relatively short period. In this chapter we try to demonstrate how to overcome some of the hurdles in achieving the goal of introducing a cost-effective biopesticide to the "world" of postharvest disease control.

II. THE MICROBIAL ANTAGONISTS

The initial step in the development of a commercial agent for biological control is fulfillment by the potential antagonist of certain criteria on the basis of laboratory studies.

Wilson and Wisniewski[2] have discussed the two basic approaches that are available for using antagonistic microorganisms. One is the suppression of disease by an epiphytic population of microorganisms that may be enhanced to resist disease expansion during storage. Alternatively, antagonists can be artificially introduced to wound sites of stored commodities and through selective manipulations can enhance a similar process of disease suppression. The latter is of interest to the industrial world.

Although criteria for an "ideal antagonist" were proposed earlier in the review of Wilson and Wisniewski,[2] they are presented here with a very strong view toward the marketing requirements. The desirable characteristics are

1. Genetic stability
2. High, consistent efficacy
3. Ability to survive under adverse environmental conditions
4. Effectiveness against a wide range of pathogens on a variety of fruits and vegetables
5. Amenability for growth on an inexpensive medium in fermentors
6. Stability of the end product during storage
7. Nonproduction of secondary metabolites that may be deleterious to humans
8. Strain resistance to standard fungicides
9. Compatibility with other chemical and physical treatments of the commodity such as heating and waxing

Among potential antagonists, yeasts deserve special attention, as is proposed by Wilson and Wisniewski[4] and as will be evident from this discussion.

Antagonistic microorganisms can come from several sources. The most adequate candidates, however, are those derived from the epiphytic microflora of the commodity to be protected. For a skillful selection of the antagonist, the "silver bullet" approach of Spurr and Knudsen[5] is probably the most promising for commercial purposes on scaling up of the process. A single antagonist rather than a mixture is a prerequisite for successful mass production and subsequent commercialization. A unidimensional process can become a multidimensional complexity which can then be reflected in the cost of production and registration. The latter has to be accounted for in any decision-making process prior to scaling up. Jutsum[6] in his review on the commercial application of biological control has stressed the fact that industry is keen to incorporate biological control agents in their programs of pest control. Among the existing microorganisms, the greatest interest is directed to bacteria followed by fungi because they are simpler to harness for mass production in the industrial setups which are available today. Moreover, despite remarkable progress that has been made in recent years to establish the basic understanding of biological control of postharvest disease, there is still a gap between the current knowledge coming from research institutions and the information on the efficacy of the various organisms as a system for pest control in the industrial environment.

The first successful event of pilot testing was that described by Pusey et al.[7] for the control of *Monilinia* brown rot on peaches. An extensive multifaceted program was initiated several years later based on the assumption that yeasts originating from the epiphytic microflora have the potential to become a commercial product that fulfills the criteria listed above for an ideal biofungicide. Initially, laboratory and pilot studies made use of a ubiquitous isolate named *Pichia guilliermondii* (US-7) which was effective for suppression of several pathogens in different crops. Its efficacy was demonstrated by Droby et al.[8] on *Penicillium* molds of citrus. Control of *Monilinia* brown rot and *Botrytis* gray mold of peaches and apples by the same yeast isolate was demonstrated by Wisniewski et al.[9] Control was demonstrated not only in an isolated situation of the laboratory but also in a pilot line that simulates a packinghouse, as discussed extensively by Hofstein et al.[10]

This yeast isolate was subsequently replaced by another yeast identified as *Candida* sp. (isolate I-182). A novel strategy for the isolation of naturally occurring antagonists was developed by Wilson et al.[1] The strategy gave rise to isolate I-182, which during the screening for new and better agents was found to be the best nonantibiotic-producing organism. Even though antibiosis was demonstrated by Weller and Cook[11] and Mazzah et al.[12] to be a very effective mechanism for fungal inhibition, it is totally unacceptable for the purpose of postharvest disease control. Isolate I-182 was selected for further development, since it fulfilled most of the ecological and functional requirements. Most important it has the strong tendency to grow rapidly at wound sites, and therefore it follows the most favorable mode of action as a commercial biopesticide, namely, competition with pathogens for nutrients and space at a wound site. Moreover, it has a clear medical record, even under opportunistic situations, which makes it a favorable candidate from a medical standpoint. Nevertheless, prior to a commercial launching of the product, it will have to undergo full-scale toxicological testing of the "end use product." Due to the unique situation of the confined and relatively isolated postharvest environment, the environmental release of microorganisms is of lesser concern to the authorities; and therefore, the required tests do not include any ecological studies other than the acute mammalian toxicological tests.

For reasons of proprietary positions, industry prefers to protect the antagonist by a patent. The discovery of US-7 and I-182 was protected by a use patent which serves as a basis for sub-license agreements between the research institutes, in this case ARO (Israel)/ARS (U.S.A.), and industry, represented by Ecogen, Inc.[13]

III. MASS PRODUCTION OF THE ORGANISM

The art of industrialization of biological pest control is first and foremost the capability to select microbial candidates that might indeed be adjusted for mass production despite their detachment from the natural habitat of growth. Moreover, the process has to be cost-effective. The art, which for practical purposes is called fermentology (namely, microbial mass production by fermentation), involves the selection of physicochemical conditions and the most suitable growth media for massive and effective cell multiplication. It is a common guideline among fermentologists that to be cost-effective, a process must be conducted in a system of submerged fermentation. The process has to be terminated within 24 to 30 h, and the growth media have to rely on industrial waste material. All of the requirements listed above have been fulfilled in the development of a production process of the yeast antagonist. Following optimization of the fermentation process it became feasible to multiply the cell population by a 1000-fold within a period of 20 h in the process of submerged fermentation. A comprehensive program directed toward the selection of useful constituents of growth media gave rise to a variety of effective and yet extremely cheap sources of energy. The different options are summarized in Table 1.

Table 1   The effect of industrial waste material on growth of the yeast *Candida* sp. (I-182) and on its capability to inhibit *in vitro* germination of *P. digitatum*

Growth medium	Cell counts		Inhibition of *Penicillium* germination
	24 h	48 h	
NYDB[a]	5×10^8	7×10^8	94
CSM[b]	9×10^7	6×10^7	—
CSL[c]	5×10^8	2×10^8	81
Soybean meal	1×10^4	2×10^3	—
Bone meal	2×10^8	3×10^8	—
Tomato juice	3×10^8	3×10^8	—
CSL + glucose	3.5×10^8	2.8×10^8	—
CSL + sucrose	6.5×10^8	1.1×10^9	—
CSL + sucrose + KCl	1.5×10^9	1.2×10^9	94
CSL + sucrose + NaCl	19×10^8	1×10^9	—
CSL + sucrose + CaCl$_2$	9×10^8	9×10^8	—
CSL + glycerine	3×10^8	3×10^8	—
CSL + soymeal	2.3×10^8	2.3×10^8	92
CSL + soymeal + sucrose	2.3×10^8	4.4×10^8	—
CSL + CSM	1×10^8	1.3×10^8	80
CSL + barley extract	1×10^8	2×10^8	—
Orange peel extract	2.7×10^7	3×10^7	89

[a] NYDB = nutrient, yeast contact, Dextrose broth; [b] CSM = cotton seed meal; [c] CSL = corn steep liquor.

One option that was found exceptionally useful is a combination of corn-steep liquor (waste extract material from the starch industry) and sucrose (which can be replaced by any other source of energy such as molasses).

A cost-effective fermentation process is a prerequisite because the product has to establish itself in the postharvest disease control industry that is commonly investing only a small proportion of the total processing cost for the purchase of fungicides. Lowering the cost of production is the approach that enables us to develop a product which becomes price competitive with standard fungicides.

IV.   QUALITY CONTROL OF THE PRODUCT

Reproducibility in performance is, by and large, the predominant requirement of a reliable product. This can be attained by fulfillment of strict guidelines of quality assurance (QA) which are the determinants for acceptability of any product in the marketplace. For QA of pesticides, the array of *in vitro* and *in vivo* efficacy tests are probably the prevalent means for evaluation. Efficacy tests in the most simplistic sense are those tests that measure the response of a system to an external deterrent signal. The same methods that are used to elucidate the mode of action of any particular organism become the key instruments in the hands of the industrialist to determine the quality of the product in every fermentation cycle. At the end of the process and upon storage of the product, each batch ought to be qualified and certified according to its performance in a QA test. A reliable potency test is therefore as important as the mass production for success in development of a commercial product.

The *in vitro* test which was originally designed for rapid screening and selection of new antagonists has been adapted by us for evaluation of fermentation batches at the end of every cycle. The *in vitro* test is based on the capability of yeast cells to inhibit germination of fungal spores on a minimal synthetic medium. The technique is simple and reproducible but lacks the required correlation with the desired result of disease control as reflected in the pilot experiments. Therefore an *in vivo* test on artificially wounded commodities has been added to the QA procedure. In this test, cell suspensions in serial dilutions are applied to the wounds with a known amount of fungal spores. The assay which was described by Wilson and Chalutz[14] and by Wisniewski et al.[15] was found to be very useful for QA. The main limitation is the time course since it takes several days to determine the rate of disease suppression. Therefore this assay can be used only as a confirmation of the results obtained in the *in vitro* tests.

Efficacy tests have to be conducted at several time points during the production process: first, upon termination of the fermentation process which was described above as a process that lasts for about 20 h; second, at cell harvesting; and last and most important, at the end of product formulation.

Quality control tests of antagonists that function by virtue of fungicidal components (i.e., metabolites) are relatively easy to monitor by chemical analysis. The same becomes complicated wherever the antagonist functions via site occupancy and/or competition for space and nutrients as well as direct attachment. These mechanisms were proposed by Droby et al.[8] in the case of postharvest control of citrus rots and by Wisniewski et al.[16] for biological control of *B. cinerea* on apples. Unlike QA systems that use chemical analysis, here the measured signals are relatively subjective and tedious. As part of the development program it becomes of utmost importance to establish simpler and quicker QA techniques because as a common routine practice of industry, every production batch has to undergo QA tests prior to its utilization in pilot plants or packinghouses.

V. PILOT TESTING OF THE BIOLOGICAL FUNGICIDE

Pilot packing lines have been designed to simulate the crucial steps of postharvest disease control procedures. The main objective of such simulation studies is the evaluation of ideal positions along the packing lines for the biological treatment that may provide a meaningful rate of disease control. The most essential guideline is, therefore, that the pilot line represents most, if not all, of the components of the commercial packing line. A successful program was conducted by Wilson and colleagues[16] since they made use of a pilot line which simulates very effectively the commercial line of deciduous crops. The period for development of a commercial product has been accelerated substantially, as was the case with US-7 and I-182, by the skillful use of such a simulator. A similar approach was adapted for the pilot testing of citrus varieties in a pilot line that was recently erected in Israel for such purposes (see Figure 1). The pilot line can be used for selection of the site where pesticides should be applied, whether before or together with the waxing process. It can also test aspects of application instruments and methodologies. The pilot lines for citrus crops (Figure 1A) and for deciduous crops (Figure 1B) have been used to compare application of systems such as dipping, drenching, and even atomization (misting) spray techniques. It is evident from our experience that selection of the best application system as well as the positioning of the applicators can contribute to the extent of disease control. The same packing lines are used for selection of the most desirable product formulations. This condition is extremely difficult to mimic in any other setup.

The most elementary step in the development process is the demonstration that yeasts exhibit disease suppression in a concentration-dependent manner. A common feature of our pilot experiments is that the unformulated cell preparation, namely, the harvested cells without any extra additives, may reduce the disease incidence by at least 50%. In the

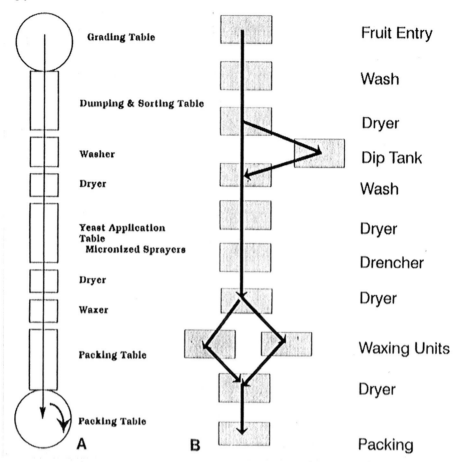

Figure 1 Schematic presentation of the pilot lines for (**A**) citrus crops (ARO, Israel) and for (**B**) deciduous crops (ARS, Kearneysville, WV).

range of 10^7 to 10^8 cells per milliliter, the expected result is attainable as is the case in a representative trial as schematically summarized in Figure 2. The latter was conducted in a situation of drenching, but it is likewise useful to apply the yeast suspension by dipping (Figure 3). The conclusions from such a series of experiments are that the required flexibility can indeed be achieved and that the concept of disease control by competition can be monitored in such pilot packing lines.

VI. PRODUCT FORMULATION

The core in any microbial pesticide is a concentrated paste of cells which are harvested following fermentation. In order for such a fresh preparation to become a commercial product it has to fulfill two major criteria: (1) induce maximal disease antagonism in variable environmental conditions; and (2) attain stability upon storage, namely, preserving performance after shelf life for several months.

Every microbial agent is selected for further development only when it successfully passes a whole series of *in vitro* and *in vivo* bioassays as mentioned above. However, on exposure of the agent to realistic situations in the field or in the packinghouse, it has to perform under much harsher environmental conditions. Therefore, the agent must be

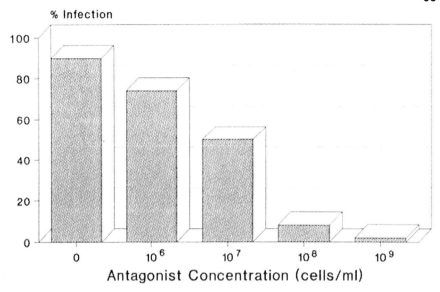

Figure 2 Percent fruit infection following drenching of grapefruit with serial dilutions of the yeast *Candida* sp. isolate I-182. The results were obtained after 2 weeks of storage at 17°C of large quantities of grapefruit (~100 kg for each dilution of yeast). It should be noted that at yeast rates of 10^7 cells per milliliter the decay is reduced by 50%.

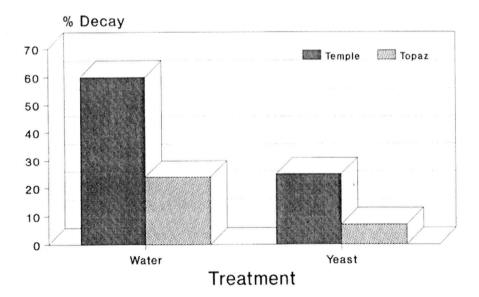

Figure 3 Percent decay of the peelable citrus varieties Topaz and Temple following dipping of the commodity in a suspension of 10^7 cells per milliliter. The check of the trial is represented by dipping of fruit in plain water. It is apparent from the results that the yeast solution supresses by 50 to 60% the rate of decay. Note that the decay is caused by natural infestation with spores of *P. digitatum*. The dip application appears to be as effective as the drench application (Figure 2).

supplemented by a variety of compounds designated "additives" or "enhancers". Among such additives may be included surfactants, humectants, adherents, and others which fortify the active agent.

Protectants are included in the formulation prior to its storage in order to avoid loss of fungicidal activity following shelf life. With the yeast, the goal is to protect the product from loss of cell viability, since the cells are expected to multiply at wound sites on application and thereby compete with the germinating spores of the pathogen. In recent years several biological products have been formulated but these were designed primarily for bioinsecticides such as *Bacillus thuringiensis* (*Bt*), whose proteinaceous crystal is the insecticidal entity. Couch and Ignoffo[17] described the principles for the formulation of insect pathogen with emphasis on *Bt* crystals. However, *Bt* products are comparatively easier to formulate than yeast since there is no requirement for viability of intact cells. It is due to the long-standing experience of our research teams that a strategy was successfully established for the formulation of intact cells. These achievements are reflected in the positive results of bioassays that are routinely used for storability. These are measuring the potential of cells to restore their growth at wound sites and indeed, despite the harsh conditions of dehydration processes, the cells remain intact.

Biological pesticides ought to be commercially distributed through chains that are commonly used for standard chemicals. Consequently not only should they be as useful as the latter, but also preferably they should be recommended for similar handling by the end user, as discussed in a review article by Rhodes.[18] The basic features of the product should be of a wettable powder or an oil-based formulation. Selection of powder carriers or oils must take into account the need to remain within the framework of ecological acceptability; namely, like the active ingredient, the carrier has to fulfill the requirements of ecological safety. The carrier is expected to create the ideal environment for the cells so that they can remain stable on storage and functional on application. The carrier elicits the protection via different modes, the predominant of which is stabilization of humidity and pH in the vicinity of the cells. Wettable powders and oil-based formulations have been designed for the yeast product. The outcome was an agent that can be handled in a fashion similar to standard fungicides in terms of shelf life and fungicidal efficacy. Conceptually, it is always correct to claim that formulation additives can only improve an already existing performance and obviously one that was monitored at least in the packing pilot line. The viability of yeast following shelf life would be implicated in the extent of pest control.

On application to the commodity, the agent has to overcome certain environmental hurdles such as extreme temperatures, lack of humidity, and biodegrading signals. It was correctly assumed by Wisniewski and Wilson[19] that the postharvest environment is much more favorable than the environment in the open field. However, for optimal performance, the microbial agent still requires protection which can be provided by adequate additives.

In all our pilot-screening tests we observed that there was correlation between cell occupancy at the wound site and the rates of disease control. Results from laboratory assays indicated that there was an insufficient cell adherence to the outer surface of processed fruit. Eliciting better attachment is a requirement that can be fulfilled through screening numerous types of stickening additives. The screening must be conducted in the pilot line because no other simulation is sufficiently representative of the actual situation in the packinghouses.

In general, the additives have to be convenient for handling and compatible with the biological agent. Of similar importance is the requirement of complying with the instructions of regulatory authorities, namely, that the additives included in any official listing document. One example is the EPA list of instructions stating that formulation constituents can be only those compounds which are cited in the Code of Federal Regulations

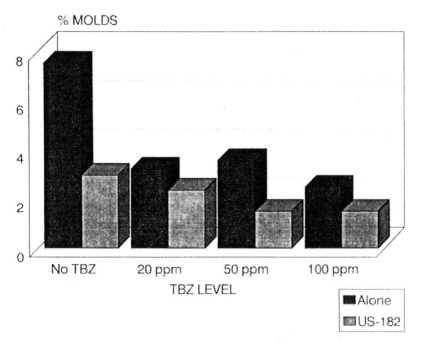

Figure 4 Dose-dependent contribution of TBZ (the range was 20 to 100 ppm, which is 1 to 5% of the recommended rate) to the disease suppression with *Candida* sp. in isolate I-182. The yeast concentration was 10^7 cells per milliliter. It is apparent that yeast as a stand-alone treatment reduces decay by 60%, and extra reduction is attained by addition of reduced rates (50 to 100 ppm) of TBZ.

(CFR, National Archives and Records Administration). At the same time, the additives must comply with the cost-effectiveness requirement.

VII. INTEGRATED PEST MANAGEMENT

An ideal product is a stand-alone agent that can suppress disease symptoms independently of its severity. However, on occasion and for a variety of reasons, the biological agent is incapable of fulfilling this requirement by itself. In situations of exceptionally high disease pressure, the alternative approach should be the combination of the biological agent with the protective agent, whereas the curative treatment should reside with standard chemicals. This strategic approach was elaborated by Eckert,[20] who proposed to control wound-invading pathogens by a series of the following treatments: (1) disinfection of the surface of fruit and its environment (i.e., SOPP, etc.), (2) eradication or suppression of germinating spores at wound sites by a combination of fungicidal agents, and (3) reduction of wound susceptibility with protective fungicides such as the biological agents. The approach of integrated pest management (IPM) was exploited recently by our research teams. The idea was to mix the biofungicide with substantially reduced amounts of various chemical fungicides such as thiabendazole (TBZ) for citrus or mertec and captan for deciduous commodities. The chemical fungicides were always applied in one tenth or less of the recommended rates.

By adapting IPM strategies, not only may we expect to gain effective pest control, but also we can maintain very low levels of chemical residues. The usefulness of this

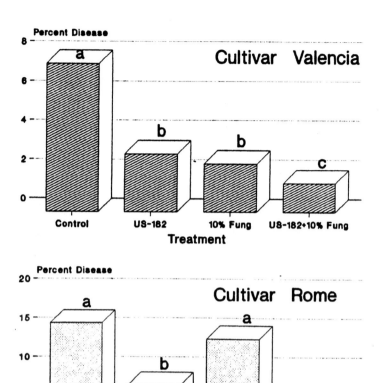

Figure 5 The effect of *Candida* sp. isolate I-182 as a stand-alone or in the presence of reduced rates of commercial fungicides, being TBZ for *P. digitatum* on citrus (ARO, Israel) and Mertec for *P. expansum* on apples (Kearneysville, WV). The commodities were treated with 10^7 yeast cells per milliliter with or without the commercial fungicides and were then stored at 17°C for 21 d.

approach was demonstrated by mixing the yeast product with one tenth the recommended rates of TBZ and mertec for the control of *P. digitatum* and *P. expansum* on citrus and apples, respectively. Typical results of such experiments are presented in Figures 4 and 5.

The results indicate that the yeast product alone can suppress *Penicillium* molds by 50 to 70%, but over 90% control is attained only in the presence of TBZ. Combination of the two is a very useful test case of IPM. For all practical purposes this should be recommended wherever severe disease pressure requires an additive effect of biological and chemical agents. It should, however, be emphasized that in these experiments, the yeast product is still devoid of the formulation additives. It is already evident from our recent studies that in many instances, and even at relatively high rates of disease incidence, the yeast product plus the additives could suppress infection without the assistance of the chemical fungicide. Obviously, the biological agent must be of low sensitivity to any of the supplemented chemical fungicides.

VIII. FUTURE PROSPECTS

It is evident from our analysis of the current status of postharvest disease control that there is no single solution to such a complex problem. It was emphasized that the transition from the laboratory to the environment of the packinghouse is not a trivial process. For success, a comprehensive strategy must be adopted. We propose to consider every existing means as long as the goal of pest control is reached while medical and ecological considerations are well preserved. This goal can be reached only by incorporation of microbial agents, together with other physicochemical treatments, into the program of pest control.

The basic assumption was that on banning of many chemical pesticides, novel and safer means will enter into the scene. Apparently, the best replacement, so far, seems to be the naturally occurring microorganisms. These, following isolation and precise characterization, were harnessed successfully to become functional as postharvest control agents. Nevertheless, it was demonstrated that in many situations, and primarily when the disease pressure is relatively high, a comprehensive IPM program should provide the expected rate of control. IPM should be considered when we are confronted by more than one pathogen, as is often the case. We discussed the various options for IPM and primarily the effectiveness of confirming the microbial antagonist together with much reduced amounts of standard chemical fungicides. We also emphasized the opportunity that resides with formulation additives. These two should not be considered contradictory. On the contrary, we ought to find ways to combine them into a comprehensive formulation. It should be stressed once more that only a reliable formulation can turn an interesting microbial antagonist into a viable commercial product. We anticipate a wave of many new and improved formulations of biological pesticides, in general, and in the arena of postharvest disease control, in particular. Once registered, these formulations will become the end use products on the standard list of recommended postharvest treatments.

The whole process can and indeed will be accelerated — regardless of all the hurdles — with the maturation of research techniques. Better and quicker bioassay techniques is one example. Fine-tuned methodologies for elucidation of the mode of action is another.

A long-term prospect is the improvement of fungicidal performance by genetic manipulation of the organisms. There are already a few examples of those improvements with *Bacillus thuringiensis* for insect control and in root colonizing *Pseudomonas fluorescens* for soilborne pathogens. With the current existing body of knowledge about the molecular features of yeasts, we can start planning the types of experiments that give rise to a more stable organism, one that expresses intrinsic elicitors of fungal antagonism.

Finally, with the growing awareness of ecological issues and with problems arising from pesticide resistance to many standard fungicides, we should expect to see an unprecedented renaissance in the demand for antagonistic yeasts in the postharvest control of fruit and vegetable disease.

REFERENCES

1. **Wilson, C. L., Wisniewski, M. E., Biles, C. L., McLaughlin, R., Chalutz, E., and Droby, S.,** Biological control of post-harvest diseases of fruits and vegetables: alternatives to synthetic fungicides, *Crop Prot.*, 10, 172, 1992.
2. **Wilson, C. L. and Wisniewski, M. E.,** Biological control of post-harvest diseases of fruits and vegetables: an emerging technology, *Annu. Rev. Phytopathol.*, 27, 425, 1989.
3. **Spurr, H. W., Vern, J. E., and Thal, W. M.,** Managing epiphytic microflora for biological control, in *Biological Control of Postharvest Diseases of Fruits and Vegetables, Workshop Proc.*, Wilson, C. L. and Chalutz, E., Eds., U.S. Department of Agriculture, Agricultural Research Service, 1990.

4. **Wilson, L. W. and Wisniewski, M. E.,** Future alternatives to synthetic fungicides for the control of post-harvest diseases, in *Biological Control of plant disease,* Tjamos, E. S., et al, Eds., Plenum Press, New York, 1992, 133.

5. **Spurr, H. W. and Knudsen, G. R.,** Biological control of leaf disease with bacteria, in *Biological Control Strategies on the Phylloplane,* Windels, C. E. and Londow, S., Eds., APS Press, St. Paul, MN, 1985, 45.

6. **Jutsum, A. R.,** Commercial application of biological control: status and prospects, *Philos. Trans. R. Soc. London,* 318, 387, 1988.

7. **Pusey, P. L., Hotchkiss, M. W., Dulmage, H. T., Baumgardner, R. A., Zehr, E. I., Reilly, C. C., and Wilson, C. L.,** Pilot tests for commercial production and application of *Bacillus subtilis* (B-3) for post-harvest control of peach brown rot, *Plant Dis.,* 72, 622, 1988.

8. **Droby, S., Chalutz, E., Wilson, C. L., and Wisniewski, M. E.,** Characterization of bio-control activity of *Debaromyces hansenii* in the control of *Penicillium digitatum* on grapefruit, *Can. J. Microbiol.,* 35, 494, 1989.

9. **Wisniewski, M. E., Biles, C., and Droby, S.,** The use of the yeast *Pichia guilliermondii* as a biocontrol agent: characterization of attachment, in *Biological Control of Postharvest Diseases of Fruits and Vegetables, Workshop Proc.,* Wilson, C. L. and Chalutz, E., Eds., ARS Publishers, 1990.

10. **Hofstein, R., Droby, S., Chalutz, E., Wilson, C. L., and Fridlender, B.,** Scaling-up the production for application of an antagonist — from basic research to R & D, in *Biological Control Postharvest Dis. Fruits Vegetables, Workshop Proc.,* Wilson, C. L. and Chalutz, E., Eds., Agricultural Research Service, U.S. Department of Agriculture, 1990.

11. **Weller, D. M. and Cook, R. J.,** Suppression of take-all of wheat by seed treatments with flourescent pseudomonads, *Phytophatology,* 73, 453, 1983.

12. **Mazzah M., Cook, P. S., Tomashov, L. S., Weller, D. M., and Pierson, L. S., III,** Contribution of phenazine antibiotic biosynthesis to the ecological competence of fluorescent pseudomonads in soil habitats, *Appl. Environ. Microbiol.,* 58, 2616, 1992.

13. **Wilson, C. L. and Chalutz, E.,** *Pichia guilliemondii* Useful for the Biological Control of Postharvest Rots in Fruits, U.S. Patent 5,041,384, 1991.

14. **Wilson, C. L. and Chalutz, E.,** Post-harvest biological control of *Penicillium* rots of citrus with antagonistic yeasts and bacteria, *Sci. Hortic.,* 40, 105, 1989.

15. **Wisniewski, M. E., Wilson, C. L., Chalutz, E., and Hershberger, W.,** Biological control of post-harvest diseases of fruit inhibition of *Botrytis* rot on apple by an antagonistic yeast; in Proc. 46th Ann. Meet. Electron Microscopy Society of America, G. W. Bailey, Ed., San Francisco Press, 1988, 290.

16. **Wisniewski, M. E., Biles, C., Droby, S., McLaughlin, R., Wilson, C. L., and Chalutz, E.,** Mode of action of the post-harvest biocontrol yeast, *Pichia guilliermondii,* Characterizations of attachment to *Botrytis cinerea, Physiol. Mol. Plant Pathol.,* 39, 245, 1991.

17. **Couch, T. L. and Ignoffo, C. M.,** Formulations of insect pathogens, in *Microbial Control of Pests and Plant Diseases, 1970–1980,* Burges, H. D., Ed., Academic Press, 1987, chap. 34.

18. **Rhodes, D. J.,** Formulation requirements for biological control agents, *Aspects Appl. Biol.,* 24, 145, 1991.

19. **Wisniewski, M. E. and Wilson, C. L.,** Biological control of post-harvest diseases of fruits and vegetables: recent advances, *Hortic. Sci.,* 27, 94, 1992.

20. **Eckert, J. W.,** Role of chemical fungicides and biological agents in post-harvest disease control, in *Biological Control of Postharvest Diseases of Fruits and Vegetables, Workshop Proc.,* Wilson, C. L. and Chalutz, E., Eds. ARS Publishers, 1990, 14.

Chapter 9

Patenting Procedures in the U.S. for Biocontrol Agents

Janelle S. Graeter and Gail E. Poulos

CONTENTS

I. INTRODUCTION

In recent years, naturally occurring biocontrol agents have become increasingly attractive alternatives to chemical pesticides. A reluctance to use chemical pesticides because of their adverse environmental impact and a growing inclination on the part of consumers to purchase pesticide-free agricultural products have combined to provide considerable impetus to the commercial development of natural products. In addition, a number of chemical pesticides, as well as some fungicides and herbicides, have been removed from use by regulatory agencies such as the Environmental Protection Agency (EPA), thereby eliminating their use entirely.

Industry has shown some reluctance to develop these products due to the financial commitment required and the financial risk involved. Patent protection, however, may offset some of those risks by providing a monopoly on a patented technology to the patent owner for a specific period of time.

In 1980, the Supreme Court opened the way to obtaining patent protection of biocontrol agents which utilize microorganisms by permitting patent protection of microorganisms for the first time. In *Diamond v. Chakrabarty,* the court held that microorganisms constituted compositions rather than products of nature. Prior to this landmark decision, microorganisms had been excluded from patent protection because, as products of nature, they fell in a nonstatutory category of invention as defined by the patent laws. Since this decision, patents have been granted on novel microorganisms for a wide variety of uses including biocontrol agents and methods.

II. PATENT APPLICATION FORMAT

A U.S. patent is an exclusive right to an invention granted to an inventor for a period of 17 years in order to promote the progress of science and the useful arts. The basis for U.S. patent and copyright laws are provided by the U.S. Constitution in Article 1, Section 8:

"The Congress shall have the power. . . to promote the progress of science and useful arts, by securing for limited times to authors and inventors the exclusive right to their respective writings and discoveries."

The purpose of a patent is to provide the inventor the right to exclude others from practicing the invention while at the same time disclosing the new invention to the public so that others would be motivated to make more advances in that particular art. A patent can be a utility, design, or plant patent. Usually, patents for biocontrol agents are utility patents.

Patent applications prosecuted before the U.S. Patent and Trademark Office (PTO) must meet requirements of utility, novelty, and nonobviousness — as set forth in 35 USC 101, 102, and 103 — in order to be deemed allowable and passed to issue. In addition, an invention must fall into one of the five statutory categories of invention: a machine, a process, a manufacture, or composition of matter, or improvements of the above. Biocontrol agents are compositions of matter, processes, or improvements thereof.

An application for a patent is made, in writing, to the Commissioner of Patents and Trademarks. All applications are required to have a specification which includes at least one claim, an oath or declaration, and drawings (when necessary). A filing fee must be paid, and all applicants are encouraged to file an information disclosure statement with the application.

The specification consists of: (1) the title of the invention; (2) the cross-references to related applications, if any; (3) a statement as to rights to invention made under federally sponsored research and development, if any; (4) a background of the invention which includes (a) field of the invention and (b) description of the prior art; (5) a summary of the invention; (6) a brief description of the drawings, if any; (7) a detailed description of the invention; (8) the claims; and (9) an abstract of the disclosure. Each of these sections, except for the title of the invention, are headings in the specification.

A. TITLE

The title of the invention is generally two to seven words and should be descriptive and technically accurate. If the title does not correspond to the claimed invention, the applicant will be requested to change the title; or at the time of allowance, the examiner may change it if a satisfactory title has not been supplied.

B. CROSS-REFERENCES TO RELATED APPLICATIONS

Cross-references to related applications occur when the application is for an invention disclosed in an application previously filed in the U.S. The application can be a continuation, a continuation-in-part (CIP), a divisional, or a reissue of a previously filed application. The cross-reference is in the first sentence of the specification.

C. STATEMENT OF RIGHTS

With respect to item 3 in Section II of this chapter, a Government contractor who retains U.S. domestic patent rights is required to include the following statement at the beginning of the application and any patents which issue thereon:

"The U.S. Government has a paid-up license in this invention and the right in limited circumstances to require the patent owner to license others on reasonable terms as provided for by the terms of Contract No. (or Grant No.) awarded by (Agency)."

D. BACKGROUND OF THE INVENTION

The background of the invention section includes two parts. The first is the field of the invention, which describes the art area of the invention. This helps the PTO assign the application to the proper examining group. The second part, description of the prior art, discloses what is already known in the relative area of the invention and includes prior use, publications, and patent disclosures. This part also includes information about any problems related to what is already known and how the present invention solves these problems.

E. SUMMARY OF THE INVENTION

In the summary of the invention section, the inventive concept is set forth in general terms. Advantages should be pointed out as well as any problems the invention solves. The summary is separate and different from the abstract of the disclosure.

F. BRIEF DESCRIPTION OF THE DRAWINGS

If drawings are included in the specification, there must be a brief description of each of the drawings.

G. DETAILED DESCRIPTION OF THE INVENTION

In the detailed description of the invention, the invention must be described so as to enable any person skilled in the art to make and use the invention without undue experimentation. The description must distinguish the invention from other inventions and from old or already known details. It must describe completely a specific embodiment and its operation. Futhermore, the best mode of carrying out the invention must be set forth.

H. CLAIMS

The detailed description of the invention is followed with a claim or claims particularly pointing out and distinctly claiming the subject matter that the applicant regards as his invention.

I. ABSTRACT

Finally, every specification concludes with an abstract of the disclosure. The abstract enables the PTO and the public to determine quickly the nature and gist of the disclosure.

III. PATENTABILITY REQUIREMENTS

A. UTILITY

As stated above, an invention must meet the statutory requirements for utility, novelty, and nonobviousness. The utility requirement is set forth in 35 USC 101 which states:

> "Whoever invents or discovers any new and useful machine, process, manufacture, or composition of matter or any new and useful improvement thereof, may obtain a patent therefore...."

Usefulness may be shown or suggested; and, if found reasonable, it is generally considered sufficient if it is disclosed in the specification or is predictable.

Closely linked with the concept of utility is that of operability, i.e., if an invention is inoperable, what would it be useful for? Generally, an inventor is not required to show proof of operability in order to demonstrate utility; however, under some circumstances, an examiner may present evidence or reasoning indicating doubt that a claimed invention would be useful for its intended purpose. In this instance, the submission of some evidence would be necessary to overcome a rejection on the basis of lack of utility.

An inventor is not required to disclose every application in which an invented product might be used, and it is not required that an invention be better than or equal to that which is already known in order to be patentable. In addition, while a showing of commercial success may be considered by an examiner, it is not a requirement for patentability.

In the instance of biocontrol agents, for example, novel microorganisms such as bacteria, which are useful biocontrol agents, may also have properties which are useful in other processes. Those processes, however, do not have to be disclosed — only those to which the invention pertains, i.e., biocontrol. In addition, some strains of a novel bacterium may show activity which is superior to that of other strains of the novel bacterium. As long as all strains show some activity, all strains would be patentable.

B. ENABLEMENT

In addition to the utility requirement is the requirement for enablement. Not only must an invention be useful, but also one "skilled in the art to which it pertains" must be able to practice it as described by the specification. The U.S. statute, 35 USC 112, first paragraph, states:

"The specification shall contain a written description of the invention, and of the manner and process of making and using it, in such full, clear, concise and exact terms as to enable any person skilled in the art to which it pertains, or with which it is most nearly connected, to make and use the same, and shall set forth the best mode contemplated by the inventor of carrying out his invention."

Thus the enablement requirement necessitates a description of the invention in sufficient detail to enable one to make and/or use it effectively, i.e., the critical factors which affect operability must be disclosed.

With respect to novel microorganisms used in biocontrol procedures, a microorganism may be enabled in two ways. It may be described in such a way that a skilled scientist would be able to obtain it in usable form from the description in the specification. For example, the source of the microorganism, its identifying characteristics, and an isolation procedure — especially if a step which is not conventional is used — would all be information which would pertain to the enablement requirement. Alternatively, a particular microorganism may be placed in a recognized depository under the terms of the Budapest Treaty and a statement made to the effect that all restrictions to the public would be irrevocably removed on the granting of a patent. It should be noted, however, that when using a deposit as a means of fulfilling the enablement requirement, the claims will be limited to the particular novel microorganism or microorganisms deposited. They will not encompass other strains or isolates which might be expected to be effective if a deposit is the only means of enablement provided. The objective of the requirement is to make the claimed microorganisms freely available to the public; and, if only the deposited microorganism has been enabled, it is the only embodiment which can be found patentable.

With respect to novel biocontrol methods, the description should be sufficient to enable one to successfully practice a particular method without undue experimentation. Including specific working examples in the specification generally provides sufficient guidelines to enable one to successfully repeat a process, and the examples have the added benefit of providing data demonstrating effectiveness and thus usefulness of the invention. In addition, if a particular method depends on the use of a novel microorganism, that microorganism must also meet the enablement requirement as discussed above.

C. NOVELTY

Not only does an invention have to be useful, but also it must be novel and unobvious. The requirement for novelty is set forth in Section 35 USC 102 which states:

"A person shall be entitled to a patent unless —

(a) the invention was known or used by others in this country, or patented or described in a printed publication in this or a foreign country; before the invention thereof by the applicant for patent, or

(b) the invention was patented or described in a printed publication in this or a foreign country or in public use or on sale in this country, more than one year prior to the date of the application for patent in the United States, or

(c) he has abandoned the invention, or

(d) the invention was first patented or caused to be patented, or was the subject of an inventor's certificate, by applicant or his legal representatives or assignee in a foreign country prior to the date of the application for patent or inventor's certificate filed more than twelve months before the filing of the application in the United States, or

(e) the invention was described in a patent granted on an application for patent by another filed in the United States before the invention thereof by applicant for patent, or on an international application by another who has fulfilled the requirements of paragraphs (1), (2), and (4) of 35 USC 371 (c) of this title before the invention thereof by the applicant for patent, or

(f) he did not himself invent the subject matter sought to be patented, or

(g) before the applicant's invention thereof the invention was made in this country by another who had not abandoned, suppressed, or concealed it. In determining priority of invention there shall be considered not only the respective dates of conception and reduction to practice of the invention, but also the reasonable diligence of one who was first to conceive and last to reduce to practice, from a time prior to conception by the other."

With respect to biocontrol agents, the agent itself and/or the method of using it may be novel. The discovery of a previously unknown property of or a new use for a known composition such as a microorganism or a pheromone, for example, does not confer patentability on the microorganism or pheromone. If the prior art discloses the biocontrol agent per se with or without a utility and enables that agent by describing it, a patent cannot be obtained on the agent — no matter what new and useful properties may have been discovered.

It is possible, however, to obtain a patent on the use of the composition if an application is submitted claiming a method and the method is sufficiently described. Only by claiming a method can the newly discovered properties be brought out. In the case of the newly identified pheromone, the compound itself may have been known, but it was not known to be a pheromone or to have potential biocontrol use. By claiming a biocontrol method, however, one can utilize its newfound property to obtain a patent.

D. NONOBVIOUSNESS

Although a composition such as a biocontrol agent may be useful and novel, it must also be nonobvious. The requirements for nonobviousness are set forth in 35 USC 103 which states:

"A patent may not be obtained though the invention is not identically disclosed or described as set forth in section 102 of this title; if the difference between the subject matter sought to be patented and the prior art are such that the subject matter as a whole would have been obvious at the time the invention was made to a person having ordinary skill in the art to which said subject matter pertains. Patentability shall not be negatived by the manner in which the invention was made. Subject matter developed by another person, which qualifies as prior art only under subsection (f) or (g) of 35 USC 102, shall not preclude patentability under this section where the subject matter and the claimed invention were, at the time the invention was made, owned by the same person or subject to an obligation or assignment to the same person."

Some criteria used to determine nonobviousness are whether the invention step(s) was (were) (1) an exercise of ordinary mechanical skill; (2) a perfection of the workmanship; (3) a logical deduction from the teachings of the prior art; (4) a carrying forward of an old idea; (5) a substitution of a known equivalent for one of the elements of a known structure, composition, or compound; or (6) a change in size, degree, or form of a known invention. Furthermore, the PTO is required to use a set of factual inquiries as a test for obviousness as set forth in *Graham v. John Deere Co.,* 383 U.S. 1, 148 USPQ 459 (1966). These factual inquiries are

"1. Determining the scope and contents of the prior art;
2. Ascertaining the differences between the prior art and the claims at issue; and
3. Resolving the level of ordinary skill in the pertinent art."

Finally, in determining the patentability of any invention, each case is determined on its own merits using the above as guidelines for the examination process.

Chapter 10

Commercialization of Biological Control Agents: An Industry Perspective

Steven K. Whitesides, Richard A. Daoust, and Richard J. Gouger

CONTENTS

I. INTRODUCTION

More than 30 years have now elapsed since Rachel Carson published *Silent Spring*, her stunning revelation of the grave environmental consequences of our reckless overuse of chemical pesticides. Her book heightened public awareness about the environment and its fragile nature. It ultimately led to public and governmental risk assessment over use of certain classes of chemical pesticides. Risk assessment led to establishment of safety policies and exposure tolerances for pesticides that resulted in banishment of high-risk chemical compounds, a trend that continues to the present. While *Silent Spring* can be credited with awakening public awareness, agricultural production still has widespread reliance on the use of chemical pesticides. This is evidenced by the fact that chemical pesticides presently still account for more than 98% of all pesticides sold in the world.

In the last few years, new and even more ominous concerns have arisen over the widespread use of chemical pesticides. The pests themselves, including damaging species of insects, plant pathogens, and noxious weeds, are becoming resistant to the very chemicals that are applied at high cost to eliminate them. In the case of insects, for example, the number of species that have become resistant to chemical insecticides has increased logarithmically over the last few decades, going from less than 10 species in the late 1940s to more than 500 species today.[1]

Even the public and print media are becoming increasingly aware of this trend as described in a recent *U.S. News and World Report* article entitled "The Joy Ride Is Over: Farmers Are Discovering That Pesticides Increasingly Don't Kill Pests."[2] If alternative pest control procedures are not found soon, the alarming trend of pest resistance could lead to increased chemical rates and application frequency. Increased application frequency can intensify selection pressure in a pest population which can perpetuate resistance buildup and eventually lead to pest control failures. This cycle of chemical dependence, overuse, increased resistance, and eventual pest control failure is not interrupted unless alternative tools are inserted. Farmers could experience catastrophic economic losses as chemical application cost inputs increase, yet harvest yields decrease due to pest control failures. In addition, consumers could face limited availability and inflated costs of agricultural goods. Consumers also want assurance that their food is safe to eat. There exists a perception that growers' dependency on chemical pesticides is cause for alarm because of health risks associated with chemical residues on food.

In no agricultural sector are these concerns more acute than among fruit growers who rely heavily on the use of chemicals for both preharvest and postharvest treatments of fruit. Public health organizations urge consumers to eat fresh fruit to ensure good health. Prolonged storage of fruit is made possible by residual chemical control of decay-causing pathogens. Ironically, the chemical residues that protect fruit are also implicated to cause undesirable health risks. Regulatory policy under consideration for residue tolerances includes a zero-risk factor for chemical exposure. Advances in measurement sensitivity for detection of chemical residue enables detection capability at extremely low levels. Sensitive decay-causing pathogens are usually controlled at low chemical rates; however, but as resistance increases, so must treatment application rates and frequency. The presence of very low, yet detectable levels of chemical residue remains a debatable health risk, and the need to elevate residues for effective decay control aggravates the dilemma. Certain residue levels may limit sales into domestic markets and prevent distribution into even less tolerant international markets.

Increased concern for our health and the environment has created the demand for innovative alternate pest control procedures. The most widely discussed alternative is the use of biological control agents in integrated pest management (IPM) practices. IPM relies on the premise that safer biological and cultural control practices should be employed first. Then only when absolutely necessary should chemicals be used.

In this chapter, we attempt to present an industrial perspective on the commercialization of biological disease control agents for use in postharvest decay control. Growers desperately need effective biological disease control products to replace standard chemicals, augment their effectiveness, or provide control where currently none exists. Industry is beginning to respond to this need as indicated by the numerous biological agents that are now being evaluated. Some of these agents have already been field tested and are showing good promise as replacements for more traditional treatments. As the pest control battle wages on, however, it is not yet entirely clear how many of these "new generation" biological control agents will actually be developed and sold by industry.

II. THE DISCOVERY AND DEVELOPMENT PROCESS

Historically, many attempts have been made to use biological agents to control insect, weed, and disease pests. These attempts have generally, but not always, arisen as an extension of observed relationships in nature. Natural coexistence of predator and prey or parasite and host is commonly found. This has often led scientists to search for potential biocontrol agents at the site of origin of their host species. With regard to pathogens, this search can even extend beyond habitats where the targeted organism is pathogenic into unique niches where the organism may exist as a saprophyte.

The biocontrol agent may not be as cosmopolitan as the host pathogen in habitat range and diversity of adaptation. This could result in concerns regarding the release of biocontrol agents into regions where they do not exist. Ideally, the association of the biocontrol agent with the host will be ubiquitous, giving more promise that the agent can provide effective control under all environmental conditions in which the host exists. If not, precautions to ensure containment of biocontrol agents may be required until risk assessments of their nonindigenous release and environmental impact on nontarget species can be conducted.

The most commonly accepted measurement for successful introduction of a biocontrol agent is its ability to control the target species at the diversity of natural habitats occupied by the target. Initial evaluations for commercial development are made under artificial conditions in a laboratory or greenhouse. Successful control in a simulated arena may not extrapolate to natural field conditions. The introduction of the biocontrol agent beyond its indigenous habitat can often bring unpredictable and disappointing results. It is now generally accepted that a biocontrol agent can provide acceptable control across many habitats only if an imbalance of the natural population dynamics between the biocontrol agent and target pathogen is created which favors the biocontrol agent. The most common means to achieve this imbalance is by introduction (inoculation) of abnormally high numbers of the biocontrol agent in a timely manner such that it can have a competitive advantage as an antagonist or parasite against the host. An understanding of the coexistence behavior of both organisms is extremely important in order to appropriately select inoculation timings to maximize competitiveness of the biocontrol agent and exploit the vulnerability of the target pathogen.

Some of the greatest successes in biological control have resulted from specific, obligate dependencies on the host species. Often the best biocontrol candidates have been selected following observations of their effectiveness under specific conditions. The requirement for specificity is an asset for defense of registration challenges but a market-limited liability when effectiveness is confined to such a restricted use. Pests can be either related taxonomically or behaviorally. The ability of a biocontrol agent to control closely related species is ideal. Alternatively, pest species can also be closely related to innocuous or beneficial species, so that crossover control is undesirable. In reality, most biocontrol agents are limited in the number of pest species they can effectively control. Also, their release rarely disrupts the ecosystem or adversely impacts the beneficial species present.

Perhaps the best example of this is the biocontrol agent, *Hirsutella thompsonii*, which has been commercialized in the United States. This agent provides effective control of several important species of plant parasitic mites, particularly eriophyid and tetranychid mites which inhabit citrus, grapes, and other crops. It is generally regarded as safe against most species of predatory mites, and no adverse effects have ever been observed against other nontarget species.[3]

It is very important to understand and characterize the biology of the biocontrol agent. Plant disease occurs as a result of a three-dimensional interaction between the host plant, the environment, and the pathogen. Traditionally, chemical control products worked well under variable stages of this interaction. Growers could delay applications using an eradicative strategy to control disease progression that was well underway. Alternatively, using a protective strategy, growers could apply chemicals to inhibit initial development of disease. Certain chemicals were effective regardless of environmental conditions, crop growth stage, or level of disease pressure.

The use of a biological control agent adds a fourth dimension to the factors leading to plant disease. Disease control provided by a biocontrol agent is dependent on how well the agent can interact and adapt with the target pathogen, host plant, and environment. We must attempt to define the limiting conditions for effective control among the interactive components the biocontrol agent will encounter. In order to achieve desirable control, the response of the biocontrol agent under controlled parameters should be characterized well enough to enable predictive responses under natural conditions. In addition, the mode of action of the biocontrol agent against the target pathogen must be sufficiently clear that the application can be made in such a way to maximize the probability for effective control.

An understanding of the dynamic interaction between the biocontrol agent, the pathogen, the host plant, and the environment is critical when trying to select optimal application procedures. Application timing, frequency, and methods should favor establishment of the biocontrol agent. The influence of formulation components, application carrier and volume, addition of adjuvants, stickers, UV protectants, buffering agents, or related tank mix components can also optimize activity of the biocontrol agent. An effort should also be made to identify inhibitory tank mix components. Spray tanks should be free of chemical residues that could inactivate the biocontrol agent. An example would be an undesirable interaction between chemical fungicide residues that could inactivate a fungal or yeastlike biocontrol agent.

As commercial development proceeds, insights into the biology and mode of action of the biocontrol agent continue to be revealed. Commonly, however, a complete understanding of specific biochemical interactions is not fully elucidated until the product is registered and put to use. Widespread use prompts expanded investigations by academic researchers to characterize specific activity. In some cases, a biochemical understanding of the mode of action is only described many years after introduction or not at all.

Biocontrol agents suppress or eliminate pathogens through either a passive or active interaction with the pathogen and/or the host. Two examples of passive interactions between a biocontrol agent and a pathogen are given. The first interaction is common in associations between epiphytes. It involves direct competition for nutrients and space. Usually this is a competition by both organisms for substrate and space. The action of one will lessen the ability of the other to colonize the same space. The difference in this case, however, is that the biocontrol agent will establish a commensal relationship with the host while the pathogen will parasitize and possibly even destroy the host. Epiphytic yeastlike organisms are currently under evaluation by our company for this exact purpose. By providing a high inoculum of our epiphytic biocontrol agent to fruit in a postharvest packinghouse, effective colonization by decay pathogens is suppressed or entirely prevented.

The second passive interaction between a biocontrol agent and a pathogen involves induction of a resistance response in the host plant. Introduction of a biocontrol agent to the plant can trigger defensive responses by the plant that prevent a pathogen from entry into or colonization on plant tissues. One type of resistance response is compartmentalization, e.g., callus production, that effectively walls off invaders. Another type would be wound-healing responses including enzyme production, changes in osmotic balance, pH, or chemical environment such that invading pathogens are discouraged from entry and colonization.

Similarly, two examples of active or aggressive interactions between biocontrol agents and pathogens are given. The first involves secretion of antibiotics or production of toxins by the biocontrol agent that are either lethal or at least discouraging to invading pathogens. Some biocontrol agents secrete one or multiple toxins that are effective biocides against many organisms including host plants and humans. This activity, while extremely effective, tends to be nonspecific and warrants extreme caution. Several species of bacteria have already been identified that produce potent toxins or antibiotics that inhibit food decay organisms, but it is still not clear whether such biocontrol agents will be effective in the postharvest treatment of fruit. More importantly, consideration should be given to whether regulatory agencies, e.g., the U.S. Environmental Protection Agency (EPA) will allow the registration of biocontrol agents that produce such biocidal compounds.

The second active interaction involves direct parasitism or predation of the pathogen by the biocontrol agent. As discussed earlier, such relationships are often found in nature and the population dynamics are such that both the biocontrol agent and pathogen populations closely cycle one another. Effective biological control relies on the introduction of high numbers of the parasite or biocontrol agent prior to the buildup of the pathogen. In the case of obligate parasitism, the pathogen must be present at levels to sustain establishment of the biocontrol parasite. This type of relationship is quite specific, and as such it is viewed as an asset for registration approval. Our company is currently developing such a specific parasite product, *Ampelomyces quisqualis*, a hyperparasite of many powdery mildew species.

Preliminary characterizations to determine specificity and aggressiveness should be ascertained. Ideally the biocontrol agent will provide control of more than one type of pathogen and yet remain innocuous to plant and animal tissues. This is a blend of nonspecificity and passiveness that will be a tremendous asset for its registration potential and commercial success.

Chemical control often provides consistent, effective control of pathogens unless or until the organism develops tolerance for or becomes resistant to the chemical. Comparable control with biocontrol agents is more difficult to manage and achieve. Generally, biocontrol agents do not replace effective chemicals as a result of performance comparisons, but rather due to the development of resistance or of safety concerns regarding use of the chemical. Safety concerns have justifiably led to more rigorous risk assessments. Chemical residue detection levels are possible at lower levels than ever before. Residual persistence of the active ingredient or secondary metabolites is often the principle method of effectiveness provided by the chemical control compound. Reduced residue allowances compromise the very potency of the chemical, that is, control by virtue of residual activity. Residual performance is now considered a registration liability. Combine these intensified challenges for registration with continued resistance development and there are an increasing number of diseases for which no chemical controls are available. By default, a biocontrol agent has a clear opportunity in these cases.

There are a variety of control strategies that integrate both chemical and biocontrol agents, previously described as IPM. Management strategies could include treatment mixtures of the biocontrol agent and a low concentration of a compatible chemical compound. Lower use rates of a chemical would result in lower residues. Another option would be to alternate applications of bio- and chemical controls. Alternate or supplemental use of biocontrol agents can be an important component of chemical resistance management. Reduced exposure of the target pest to the chemical will decrease resistance selection pressure and lengthen the useful effectiveness of the chemical. Integrated use of biocontrol agents should be more palatable to those who voice concerns about chemical dependency.

Candidate biocontrol agents must be selected very carefully from the complex environment in which they exist. The decision to initiate commercial development of a microbial pesticide must be made early in the developmental process and depends on the specific attributes of the organism. The biocontrol agent must be biologically characterized enough to determine commercial feasibility. Otherwise, it is unlikely to have a successful introduction into the marketplace. Some of the most important considerations are as follows:

- Can the microbial agent be easily cultured, and if so, is there potential to mass produce it at a cost that is competitive with current products?
- Is the microbial agent genetically stable, both before and during scaleup production, and biologically stable with regard to its ability to suppress or actively parasitize the target species?
- Can effective quality control procedures be developed that will allow for the standardization of product efficacy?
- Is the market potential high enough to justify development costs that can easily run into millions of dollars?
- Can the microbial agent itself or its use be patented? If not, can the developed product be protected from potential competitors through trade secrets?
- Is it unlikely that resistance will quickly develop, thereby rendering the product obsolete within a few years?

- Does the product possess an adequate formulation to ensure shelf life and field stability and make it easy to use and apply?
- Is the microbial agent safe to nontarget organisms as shown by an extensive risk assessment conducted to obtain registration?

With regard to postharvest decay biocontrol agents, most of these questions will require an affirmative response if the agent is to be developed commercially. It is unlikely that any company will pursue a registration for a biocontrol agent that: (1) does not pass initial Tier I toxicology tests or that causes significant problems to nontarget organisms; (2) is not easily produced at a competitive cost with existing control remedies; (3) is not stable for at least 12 to 18 months under moderate storage conditions; (4) has any further evidence that resistance may develop; (5) is not genetically stable and does not provide consistent, acceptable field efficacy; and (6) does not have as its basis a proprietary technology or patentable position. Such technologies usually require that the microbial agent can be produced through fermentation at low cost. Although this latter point may seem obvious, one of the major reasons for elimination of microbial control candidates is their inability to grow well or produce a highly stable reproductive stage during the fermentation process. A final, but equally important consideration, is that the market potential for the product must justify the high financial input that will be made so that a profit can eventually be realized. The lag time for the realization of a profit is variable and depends on the size and aggressiveness of the company and the overall investment climate.

III. PATENTS AND LICENSING OF THE TECHNOLOGY

Since a separate chapter in this text describes patent procedures in the United States for biocontrol agents, we will not attempt to review this topic in detail here. However, from an industry perspective, several approaches in addition to filing for a patent can be taken to protect proprietary biocontrol agents or technologies.

First, it is very important to select a biocontrol agent that not only is highly efficacious, but also possesses unique characteristics. The characteristics most often exploited for biocontrol agents include culturability and stability. Novel microbial strains or new species that will grow under unique conditions or on unique substrates may be sought. If high yields can be obtained at a relatively low cost, the manufacturer can guard the production system as a trade secret without competitive encroachment using the same process. Of course, if the strain can be patented as well, the company will obtain an even stronger position. Formulation technology can also contribute significantly to the development of an effective, high-quality product that is difficult to reproduce from less knowledgeable competition.

The active ingredient of a biological pesticide is generally the microorganism itself, except in the case of antibiotic or toxin production during fermentation or in association with the pathogen and host plant. In either case, formulation ingredients or tank mix additives can enhance field performance. Additives can provide physical protection of, or physical reaction with, the biocontrol agent or biocidal product. They may subtly or momentarily alter physical parameters of the microniche interface between host plant, target pathogen, and biocontrol agent to favor establishment of the latter. Additives may also trigger a host plant resistance response as described earlier that provides a colonization advantage for the biocontrol agent.

Such interactions can generally be kept as trade secrets during the developmental process. This kind of invaluable, yet unpatentable information provides considerable advantage to the originator of the technology. Pioneering research and development expertise can permit a company to enter the marketplace first and obtain an established reputation with the new product prior to challenge by a competitor with a similar product.

The licensing of a biocontrol technology or of microbial agents themselves is also possible. Many complexities are involved in negotiation between two entities including license fees, royalties on the sales of the final product, and exclusive marketing rights. It is not uncommon for the company which develops a product to license the product for distribution and sales by another company. Justification for this strategy could include the fact that the licensee may have a superior distribution and sales network in a geographic region where the licensor has no presence. This is a common arrangement for international distribution.

IV. MARKET ASSESSMENT

Among the most critical factors for determination of whether a biocontrol product should be developed is its market potential. Prior to the transition from a laboratory curiosity to a full-scale development program, a market assessment must be made to determine potential value of the product. This is especially important for biocontrol agents because developmental costs are not strictly related to market size. Instead, many costs are fixed, such as risk assessment for registration and costs associated with development of an acceptable production process. Final justification for commercial development is best determined once market assessments are completed and product value is estimated. The technical success of a product alone is not usually sufficient for development. A reasonable return on resource investment must be attainable in a reasonable period of time in order to proceed with development.

The determination of product value is linked directly to a number of parameters. These include the economic importance of the target pest species, the economic value of the crops protected, the potential acreage treated or tonnage of processed produce treated, and the costs associated with current or alternative protection. These parameters are then compared to projected costs associated with production. A basic economic profile of the target pest impact on its range of host crops should be developed. Statistics needed to generate the economic profile such as crop value, acreage or tonnage, estimates of crop loss inflicted by the target pest species, and current sales of registered control products can generally be found for each geographic region in state agricultural economic and extension documents. Commodity groups also produce similar profiles.

A comparison of competitive standards and an understanding of their strengths and weaknesses can be helpful. While the currently available standards for pest control are often chemical products, their cost to the grower is a useful measure of how much the market will support. Certainly product performance also plays a role in acceptability of a product. More value can be attached to a product that is perceived as superior to existing control remedies. If, on the other hand, product performance is only equivalent to current standards, then it will be difficult to justify higher pricing for a new product. Some value can also be added to the inherent safety of biocontrol agents, but growers and processors are generally not willing to pay much extra for this attribute. In the future, however, public concern over pesticide safety and chemical residues on food may result in more value attached to safer products.

As a final stage of market assessment, all parameters relating to commercially accepted standard products must be compared and analyzed to produce an estimated value and acceptability profile. The biocontrol agent under development must then be compared to the best products available and then assigned a value. This value must be in line with cost-of-production projections to determine whether adequate profit margins are attainable.

V. RESOURCES AND THE DEVELOPMENT PROCESS

The primary resource for the development of a biocontrol agent is trained manpower to address the various facets of developing a successful product. In the case of a biocontrol agent designed for postharvest disease control, the process includes numerous phases of development.

The discovery phase relies on the selection of a microbial agent with high potential for controlling the target pathogen as earlier discussed. The microbial species must be carefully selected before any significant development efforts are initiated. A change in species at a later time will result in the need to repeat much of the developmental work and risk assessment required for registration. This could result in delayed market introduction of the product and the potential loss of thousands, if not millions of dollars, depending on how far along the developmental process the product was at the time of the change.

In the arena of postharvest disease control, the most likely commercial product candidates for biocontrol of decay-causing pathogens are bacterial and fungal agents, generally common epiphytes found on fruits. Quite often the selection of promising candidates has been empirical. Biocontrol candidates are commonly isolated from fruit in direct association with pathogens. A higher population of nonpathogenic organisms is often associated with decay suppression. Following these observations, bioassays are designed to determine interactive antagonisms between selected nonpathogenic epiphytes and decay-causing pathogens. Initial evaluations may be as simple as growth inhibition bioassays on an agar-solidified medium.

The next aspect of screening can involve evaluation of promising candidates in fruit-dip bioassays or more elaborate trials in small-scale, pilot-processing lines. Treatments using fruit as the substrate for evaluation of decay control can include many variables. Examples of these include challenge inoculation of the pathogen before or after introduction of the biocontrol candidate, treatments with wounded and nonwounded fruit, variation in posttreatment temperature regimes, combination treatments between biocontrol candidates and adjuvants, chemical standards, and other biocidal treatments such as radiation or heat.

Generally, losses to decay in high-quality fruit are limited to less than 5 to 10% so that inoculation of decay-causing pathogens is useful to help determine statistical significance between treatments. Fruit wounding to cause tissue trauma can also result in higher levels of decay and provide a more critical evaluation of a biocontrol candidate. Postprocessing temperature storage conditions can vary between fruit types or similar fruit of varied maturity and integrity; therefore an evaluation of posttreatment temperature regimes is an important screening process for biocontrol candidates. Storage temperatures are selected to maximize storage life of fruit and, at the same time, minimize decay. A biocontrol agent that effectively grows under cold storage temperatures may have a definite advantage over a pathogen whose growth is suppressed under those temperatures. Commercial success of biocontrol agents may require modified fruit handling and storage practices. They may not fit as decay control agents by direct substitution for chemical controls in the same manner chemicals are currently used.

Once a biocontrol candidate has been selected, the second phase of development begins, product characterization. This phase usually requires resources for trained personnel and sophisticated equipment to study the biology and describe the mode of action. Initial studies are often conducted *in vitro* prior to *in vivo* evaluations designed to simulate commercial use. Sufficient evaluations should be made in order to characterize the antagonistic response of the biocontrol agent and elicit the response in a consistent manner.

An important component of product characterization is an understanding of nutritional requirements for growth. A commercial biocontrol candidate must be amenable to large-scale production. The most common production method involves fermentation. Fermentation parameters in small-scale laboratory equipment, such as shake flasks or 2 to 10 l fermentors, can be used to develop profiles of the growth characteristics of the microbial product. Fermentation microbiologists often play a central role in this phase, with most effort generated toward optimization of the production of higher yields of infective units of the product. A second objective is to retain or enhance effectiveness of the biocontrol agent following batch production. The microbiologists try to maintain the desirable antagonistic traits of the agent despite artificial production.

In vivo bioassays must be developed as part of product characterization for two important purposes. First, the fermentation microbiologist must have a way to assess the inherent effectiveness of the agent during production processes. A germination test or similar *in vitro* assay alone is meaningless unless it can be correlated to an *in vivo* bioassay that consistently indicates a desirable antagonistic response. Evaluations of virulence, host plant colonization, or direct growth inhibition of the target pathogen are examples of useful tests that can be highly correlated to an *in vitro* assay that may be much quicker to conduct and therefore are preferred as a postprocess evalution.

Indicator microbial species may be substituted for difficult-to-culture target pathogens in *in vivo* evaluation assays. Certain target pathogens may be extremely difficult to culture in contrast to an indicator species that can be easily cultured to provide a reliable index of desired antagonistic activity. A fermentation batch that consistently results in 50% germination may be acceptable if the living infective units, e.g., cells or colony-forming units, provide consistent decay control at a specific application rate. In contrast, certain influences of fermentation may weaken competitive ability of the biocontrol agent such that 95% germination results in inconsistent antagonism when competing for limited substrates in the nutrient-deficient plant host environment.

The second purpose for an *in vivo* bioassay is to establish a reproducible method to evaluate efficacy of the biocontrol product for quality control purposes. This is generally the duty of a specialized unit or department with the sole objective of monitoring product efficacy and overall product quality. A strong quality control procedure based on an *in vivo* bioassay is not only critical for ensuring that a quality product be used in field trial research projects, but also for implementing a system for quality control of the final commercial product. In the latter phases of product development, the quality control team will also need to develop other monitoring procedures unique to biocontrol agents. This will include methods to assess microbial contaminants, particularly human and animal pathogens that could contaminate the final product.

A molecular geneticist may also play a key role in further development of the biocontrol agent. As genetic characterization reveals an understanding of behavior, gene modifications may be considered to improve performance. It is very important, however, that research aimed at genetic improvement of a strain be linked directly with the growth and/or fermentation characteristics of the strain fairly early in the research program. Selection for commercial development will be based on the fermentative capabilities of the strain as much as on the genetic superiority of the strain for traits other than enhanced growth characteristics or higher yields of the active ingredient.

A profile of the genotype is desirable for at least three reasons. First, patent or license protection can be strengthened by genomic characterization. Closely related yet different strains may not be as efficacious for decay control, thereby providing the licensed strain added value for uniqueness. Second, an additional quality control measurement is provided if the strain can be profiled using a deoxyribonucleic acid (DNA) "fingerprint". Third, the ability to track strains released into the field is often a requirement of environmental fate studies. Either antibiotic resistance markers or a plasmid profile can provide definitive identification of released microbes.

Genetic techniques aimed at strain improvement are already being widely applied in the development of bacterial biocontrol agents, e.g., *Bacillus thuringiensis*. However, progress has been slower from a commercial standpoint with other agents such as fungi, viruses, protozoa, and nematodes. Considerable research in both the public and private sectors is underway, however, with many biocontrol agents to improve through genetic manipulation characteristics that could ultimately lead to useful commercial products.

VI. FERMENTATION AND DOWNSTREAM PROCESSING

As described earlier, the first two phases of commercial development are aimed at candidate selection and development to provide sufficient information about candidate biology to produce a stable and effective product. The inherent effectiveness of the biocontrol candidate should also be quantified and exploited through genetic and biological manipulations in these early stages of development.

Next the biocontrol agent enters a very complex and costly phase of commercial development, that is, fermentation and the downstream processing that follows. These processes should produce a stable, high-quality product for commercial use at a competitive cost. Ideally, a successful fermentation process not only will be established for production of experimental batches, but also it should be very close to the final process used for commercial development. The experimental batches will be used for formulation development and efficacy evaluations.

Production parameters should be clearly identified during this phase of development. Liquid fermentation, often referred to as submerged fermentation, is the most common and available method of production for biocontrol agents. At present, nearly all commercial biocontrol agents are produced using deep-tank liquid fermentors, even though cost of production by this method is expensive. The method is widely used, however, due to the ability during fermentation to carefully control and monitor the physical parameters, such as temperature, aeration, and pH. Ingredients of the fermentation medium can also be easily adjusted before or during the actual process in order to optimize nutritional and physiological requirements of the biocontrol agent. This should ultimately lead to a stable, high-quality product that will result in consistent, predictable commercial performance.

The fermentation process is initiated from a pure seed culture of the biocontrol agent. Stable culture storage techniques such as liquid nitrogen, freeze-dried, or silica gel storage should be developed so that culture viability and purity are preserved. At least two culture collection repositories should retain samples including one internationally recognized culture collection. The stable agent is transferred from storage culture into either a test tube slant or a petri dish containing an agar-solidified medium suitable for growth of the dormant cells. Once culture purity is confirmed by quality control, then a single colony is used to inoculate a seed flask containing liquid medium. After growth for a defined interval, the seed culture is used to inoculate the pilot tank. The length of time for completion of the fermentation will depend directly on the type of agent being grown, but does not generally exceed 36 h for most bacterial and yeastlike biocontrol agents. Fungi and nematodes require a considerably longer fermentation period.

The product of fermentation, beer, is concentrated and then usually spray dried prior to being formulated into the final biocontrol product. Beer concentrates are obtained either by filtration to remove most of the liquid or, more commonly, through centrifugation. Typically, it is important to treat the concentrate with a chemical agent to prevent contamination. A quality assessment is again made of the product at this stage to check for contamination. A contaminate could lead to deterioration of the biocontrol agent and require complete, costly disposal of the beer. Concentration of the beer and drying of the concentrate to produce a technical powder — i.e., a dried powder that contains the viable cells of the biocontrol agent, extracellular excretions of the cells produced in the fermenter, and unspent fermentation solids — constitute the downstream processing of the biocontrol product.

Various types of equipment such as spray dryers and fluidized bed dryers are commercially available to dry the fermentation concentrates. Factors such as impact on viability of the biocontrol agent and cost are evaluated when considering which drying process is most appropriate. The drying process can render an active agent harmless or even destroy fragile agents. Therefore, the process that permits retention of product viability and potency of the infective units is chosen even if it results in higher costs.

The technical powder (TP), although it contains a variety of undefined fermentation artifacts as described above, is generally considered pure at this stage. Compared to a chemical produced through a synthetic process, the TP contains far more impurities and is more variable from batch to batch. For that reason, it is critical that the production process be tailored to the specific agent and be carefully controlled. High yields and product quality must be consistently reproducible to ensure a cost competitive, commercial product. Although certain ingredients may be added during the fermentation or drying process to stabilize or enhance the qualities of the biocontrol agent, this is generally the role of the formulator.

VII. FORMULATION AND SHELF LIFE

By the time the biocontrol agent reaches the formulation team, the TP should be in a stable form. It is now the role of the formulators to develop a final formulation that has good handling properties and an acceptable shelf life. Ideally, formulators will attempt to develop a final product with several important attributes. These attributes include (1) formulation flexibility for different market needs; (2) miscibility and sprayability for application ease with no undesirable formulation residues; (3) acceptable shelf life at a range of storage temperatures and conditions; (4) caution label for user and environmental safety; and (5) functional and attractive packaging for storage, protection, disposal, and ease of sales promotion.

Selection of the type of formulation will depend as much on market acceptability as on the physical nature of the TP. This can vary considerably from market to market or even within similar markets in different geographic regions. In general, biocontrol agents have been formulated in both dry and liquid forms, similar to the formulation types used for synthetic chemicals. Examples of dry formulations include wettable powders, dry flowables, dusts, and granulars. Liquid formulations include aqueous- or oil-based flowables.

Numerous additives will be used to provide the product with desirable properties. These may include inert carriers or diluents, emulsifying agents, wetting agents, humectants, anticaking agents, ultraviolet screens, and nutritional components. Proper selection of additives should be made to enhance storability and efficacy. The physical and biological characteristics of the biocontrol agent may limit options. For example, a biocontrol agent may lose long-term stability following exposure to temperature extremes for prolonged periods or following rehydration. A requirement for stable shelf life of an agent may be dehydration-induced dormancy. This type of agent may not retain viability after imbibing water or be compatible with oils; therefore it could not be formulated in liquids. A final formulation vulnerable to deterioration by exposure to high humidity could successfully overcome this limitation by being packaged in a water-impermeable container.

Ideally, the formulated product will remain viable under variable storage temperatures. Prolonged exposure to high heat often results in product inactivation. In contrast, a product would have limited utility if it required refrigeration and/or had a shelf life under 3 months. Production and distribution of such a fragile product would be troublesome and may not be commercially viable. The formulated biocontrol product should also retain environmental safety as expected for a natural control. Eventual formulation types should preserve this safety by achieving registration classification of "CAUTION" as the label signal word. Environmental concerns are also addressed if the product packaging is selected for ease of disposal or recycled use.

Formulation of a biocontrol product targeted for exclusive use in postharvest process-ing may not involve as rigorous a screening process as formulation of a preharvest pesticide. The packinghouse environment should provide shielded use of the product to avoid direct ultraviolet sunlight or severe temperature fluxuations. Product storage may occur, however, in nearby pesticide storage sheds so that storage stability should not be overlooked.

The formulation process can be extremely frustrating if biological or physical charac-teristics of the TP limit options. This process involves blending additives and adjuvants into the formulation that potentially are either beneficial or detrimental to the biocontrol agent. These influences must be characterized for each agent. The unknown nature of the impurities in the TP intensifies the challenge to determine the impact of additives. Either deleterious interactions between additives and the agent itself or interactions between additives and impurities in the TP can result in loss of viability. The formulation team will need to spend considerable amounts of time early in this phase of development to determine the effect of formulation on both biological stability and physical properties of the TP. Abrupt changes in formulation strategies are extremely costly once the product is submitted for registration. An official Confidential Statement of Formula is submitted as part of the registration package. Altered formulations require resubmission with accompanying risk assessments that are prohibitive in terms of high cost and lengthy timetables for approval.

VIII. PRODUCT EFFICACY

The main objectives of efficacy evaluations are to determine the spectrum of activity and effectiveness of the biocontrol product under a variety of conditions. Initially, conditions established in primary screening procedures take place in laboratory bioassays and greenhouse trials. In either case, influence by extraneous variables are minimized or controlled. This stage of efficacy evaluation is necessary to identify the potential value of the biocontrol agent. The secondary screening stage shifts from laboratory/greenhouse to evaluations under field conditions or in the case of a postharvest decay control candidate, a pilot line or a full-scale processing line.

The locations selected for field evaluations should be representative of the geography where targeted commodities are of economic importance as determined by the market assessment. In the case of a product designated for postharvest use, fruit processing facilities will be the location of the "field tests". Rigorous field evaluations can be a challenge unless processing plants are cooperative enough to permit access to their facilities. Access to commercial lines is not readily available or practical during fruit processing, their peak use period. The next best substitute is a simulated, scaled-down version of a processing line, typically referred to as a pilot line. However, commercial use conditions must be evaluated to validate product potential. During the latter stages of product development, cooperative evaluations between commercial processing opera-tions and the product development company are advised.

There exists as great a diversity of postharvest treatment conditions as there are commodities and decay pathogens. Many criteria must be evaluated before a biocontrol agent can successfully be used for decay control. The biocontrol agent must be screened for resistance or tolerance to chemicals commonly used for both preharvest and postharvest control of fungal and bacterial diseases. Applications made prior to harvest often leave residues that provide postharvest decay control. Resistance to chemicals applied postharvest for control of physiological disorders such as scald and for waxes used as respiration inhibitors should also be evaluated.

An analysis should be completed to determine where, within the operation of commer-cial handling systems, the biocontrol agent can best be used. Possible use could occur in dump tank drenches, flumes, rinse operations, line-spray applicators and brush operations, wax applications, and/or drying tunnels. Packinghouse management may suggest preferred methods of use; and if activity of the biocontrol agent permits, these methods should be evaluated for feasibility. In addition, the agent must be evaluated for efficacy under temperatures and controlled atmospheres commonly used for storage.

Efficacy data generated for the above criteria may be necessary for effective decay control within a single variety of a commodity group. All or part of the same information may need to be evaluated for many varieties of fruits and vegetables. As an example, there are different preliminary handling requirements and temperatures between citrus varieties or varying storage durations between varieties of pears. Apples are usually stored at -1 to 4° C and can be stored up to 9 months in controlled atmosphere (CA). CA storage subjects fruit to low oxygen and high carbon dioxide levels. Decay pathogens vary in ability to grow under such conditions including different temperature and controlled atmosphere regimes. Also, fruit varieties vary in resistance to decay pathogens. It may not be possible for a single biocontrol agent to effectively control many decay pathogens under all conditions where decay occurs.

Different strategies for decay control by the biocontrol agent should be evaluated. Examples would include treatment mixtures with reduced rates of chemical pesticides, efficacy enhancement additives, or pretreatment with a general biocide such as heat or radiation. Continued long-term storage may require multiple treatments of fruit over time to recharge antagonistic colonization by the biocontrol agent. Treatment using a blend of complementary biocontrol agents could successfully control a broader spectrum of decay pathogens.

Many fruits are treated with drenches using large volumes of water. This can occur while the fruit is still in bins and even still loaded on trucks. Certain chemical treatments permit recirculating drench solutions as long as routine assays are taken so that effective parts per million (ppm) of the active ingredient is regulated. As needed recharges of the chemical maintain adequate ppm to compensate for loss in activity over time. Without the presence of a biocidal treatment in this system, pathogen populations could increase to levels that would overwhelm a competitive antagonist. To offset such a potential development, application of considerable volumes of the biocontrol agent would be necessary. This may not be the most cost-effective use of the product.

Fungicides are often applied to fruit as a mixture with wax in a line spray. This type of system can either be recirculating or nonrecirculating. In either case it is a more efficient application method, and smaller volumes of chemical are used in comparison with drench systems. Wet fruits are then dried in a heat tunnel. A biocontrol agent could best fit into this control strategy if it is compatible with wax and heat treatments.

Many biological and physical factors must be considered when evaluating biocontrol agents for efficacy. The postharvest industry currently uses broad spectrum fungicides that are effective in a wide range of postharvest handling and storage systems. Direct replacement of chemicals by a biocontrol agent without a compromise in performance or decay control is unlikely. Yet the postharvest industry may have no choice because current chemical standards are threatened to be discontinued due to reregistration challenges. Both the fruit processing and the pesticide industries would be best served by collaborating to design efficacy evaluations leading to development of effective decay control products.

IX. REGISTRATION ISSUES

In an earlier discussion, criteria were described for the selection process of a biological control candidate. Characteristics such as obligate parasitism, low toxicity, and unique modes of action are assets of specificity that limit market potential but should minimize registration requirements, thereby enhancing profitability.

Table 1   **Microbial pesticide data requirements**

Mammalian studies — Tier I
  Acute oral toxicity
  Acute pulmonary toxicity/pathogenicity
  Acute intravenous toxicity/pathogenicity
  Acute dermal toxicity
  Primary eye irritation/infection
  Hypersensitivity incidents
Nontarget organisms — Tier I
  Avian oral pathogenicity/toxicity (2 species)
  Freshwater fish toxicity/pathogenicity
  Freshwater aquatic invertebrate toxicity/pathogenicity
  Nontarget plant studies
  Nontarget isect toxicity/pathogenicity
  Honeybee toxicity/pathogenicity

The Environmental Protection Agency (EPA) has a mandate to register safer alternatives to the current chemical standards. Representatives of the EPA and industry are separately and jointly pursuing policy modifications that would streamline microbial registration requirements. Each year applications for microbial registration increase. In 1991, California officials reported that biological controls dominated applications for new pesticide registration.

Much of the policy established for chemical registration candidates is unnecessary for biological control agents. For example, residue tolerances are required of pesticides that may appear in foods, but microbial pesticides can be granted an exemption from the requirement of tolerance in the Code of Federal Regulations (CFR), part 180. As long as data are supplied that indicate the pesticide in any amount will be safe when used on food or feed crops, an exemption is granted.

Registration of biological pesticides is defined under the CFR; a specific subdivision of Guidelines for Registering Pesticides in the United States, Subpart M (Microbial and Biochemical Pesticides) provides supplemental policies. Pesticide registration candidates generally require a complex battery of studies which includes three-tier toxicology evaluations and four-tier environmental evaluations. Fortunately, registration requirements for many microbial pesticides are less rigorous and limited to Tier I acute toxicity and environmental testing. Economic considerations for a commitment to commercialize a biological control product are favored by less complex data requirements and review. Consumer interest alone cannot be the determining factor for commercial commitment.

The development of a chemical pesticide can take 6 to 10 years and cost $30 to $50 million, whereas development of a biological pesticide currently takes 3 to 5 years at a cost of $3 to $5 million. A benefit for development of a microbial pesticide for exclusive postharvest use is fewer data requirements than if the market included preharvest use. Table 1 illustrates Tier I data requirements which includes both mammalian studies and nontarget organism studies. A microbial pesticide targeted for use in the contained environment of commodity processing plants requires data to be generated from mammalian studies only. Since field use will not occur, nontarget organisms are not exposed to the product.

X. COMMERCIAL SALES

A strategy for distribution and sales of a biocontrol agent targeted for pest control use is developed simultaneously with the product. The results of efficacy evaluations help determine uses and value. Product price and marketing schemes are generally established prior to registration approval. This enables product information to be disseminated to preview anticipated use and allows a strategic time to be set for a "market launch" implementation. Markets are prioritized and selected to direct resource allocation in preparation for the product launch. A well-trained sales staff will initiate contact prior to the launch to educate customers about product use and serve notice about product availability, ordering information, and billing terms. The sales staff will review relevant efficacy data to illustrate the effectiveness of the biocontrol product. This process is not unlike preparation to launch any type of pesticide except the sales staff can highlight the value of a biological control and its associated environmental and user safety.

Commercial sales of a biocontrol product targeted for use in postharvest are uniquely different from sales of general use pesticides. A unique distribution system for pesticides exists within the packinghouse industry that needs to be understood. The current distribution network should be evaluated for use of an atypical pesticide in this arena, a biocontrol product. Once a product distribution strategy is finalized and commercial production quotas are established, then shipment dates and arrangements can be set. The limited number of packinghouses probably minimizes the need for centralized distribution warehouses. Packinghouses can be supplied directly from the manufacturer.

XI. SUMMARY

Development of a biological pesticide product is very similar to development of a chemical pesticide product in fundamental phases. Each goes through a discovery and developmental process that involves early discernment of activity and potential, proprietary protection, market assessment, resource commitment, small-scale production, toxicology tests, formulations, and efficacy evaluations; and then on to larger scale production and commercial sales preparation once it is championed to be a commercially hopeful product. Actual production differs as chemicals are synthesized and biological agents are typically fermented. Perhaps the greatest distinction between the two which influences all aspects of development is that the biological control agent is a living organism. This presents unique challenges for development, production, and use in contrast to traditional chemical pesticides, especially their use patterns. Successful performance and acceptance by growers will require technical education to expand their understanding of the unique dynamic interactions best suited for effective use of a biological control product.

REFERENCES

1. **Georghiou, G. P.**, in *Pesticide Resistance: Strategies and Tactics for Management,* National Academy Press, Washington, D.C., 1986, 14.
2. **Holmes, R.**, The joy ride is over farmers are discovering that pesticides increasingly don't kill pests, in *U.S. News World Rep.,* September 14, 1992, 73.
3. **McCoy, C. W.**, Pest control by the fungus *Hirsutella thompsonii*, in *Microbial Control of Pests and Plant Diseases 1970–1980*, Burges, H.D., Ed., Academic Press, London, 1981, 499.

Chapter 11

Commercialization, Facilitation, and Implementation of Biological Control Agents: A Government Perspective

Michael Mendelsohn, Ernest Delfosse, Carl Grable, John Kough, David Bays, and Phillip Hutton

The purpose of this chapter is to provide a summary regarding the current U.S. Environmental Protection Agency (EPA) and U.S. Department of Agriculture, Animal Plant Health Inspection Service (USDA/APHIS) oversight regarding biological control agents. Requirements for small-scale testing through full commercial use from the EPA as well as a description of the USDA/APHIS Biological Control Policy are presented.

CONTENTS

I. INTRODUCTION

The primary federal agencies involved in the regulation of biological control agents are the EPA and the USDA/APHIS. Other involved federal agencies include (1) the U.S. Department of the Interior Fish and Wildlife Service, which is involved when formal consultation by either the USDA/APHIS or the EPA is necessary regarding threatened or endangered species; (2) the Department of Health and Human Service Food and Drug Administration, which is responsible for the monitoring, compliance with, and enforcement of established pesticide tolerances (i.e., maximum legal residue levels); and (3) the USDA Food Safety Inspection Service, which authorizes proprietary substances and nonfood compounds (including pesticides) for use under USDA inspection and grading programs. In the sections that follow, the role of EPA and USDA/APHIS are discussed.

II. U.S. ENVIRONMENTAL PROTECTION AGENCY

Biological control agents are pesticides under the broad definition of the term as stated in the Federal Insecticide, Fungicide, and Rodenticide Act (FIFRA). As defined in Section 2(u), a pesticide is "(1) any substance or mixture of substances intended for preventing, destroying, repelling, or mitigating any pest, and (2) any substance or mixture of substances intended for use as a plant regulator, defoliant or desiccant" except those articles considered to be new animal drugs or animal feeds bearing or containing a new animals drug. However, the Code of Federal Regulations Title 40 (40 CFR) Part 152.20 exempts certain biological control agents from all requirements of FIFRA under section 25(b) of FIFRA since the EPA has determined that these agents are adequately regulated by another federal agency. Those biological control agents which are not exempted are (1) eukaryotic microorganisms, including protozoa, algae, and fungi; (2) prokaryotic microorganisms, including bacteria; and; (3) viruses. All of these agents are considered microbial pesticides.

Regardless of whether biological control agents are exempt from FIFRA, they are not exempt from any other applicable federal or state laws or regulations, e.g., the Federal Food, Drug, and Cosmetic Act (FFDCA). If the use of such agents results in residues in (1) food or feed or (2) animals used for food or feed, then that food or feed would be considered adulterated under Section 402[3] of the FFDCA and subject to seizure by FDA unless the residue was covered by a tolerance or exemption from tolerance, and/or food or feed additive regulations.

Development of microbial pesticides for commercial use generally would progress from the initial isolation and identification of the organism to laboratory and greenhouse testing, then to small-scale field testing, followed by large-scale field trials, and finally to registration of a commercial product. Under current regulations the EPA does not become involved in the initial stages of isolation and laboratory or greenhouse testing. However, EPA approval is required for all large-scale field tests of regulated types of biological control agents and, in certain cases, the EPA must be notified prior to an intended small-scale field test.

The first microbial pesticide registered under FIFRA was *Bacillus popilliae* in 1948. At this writing, 27 microbial pesticide active ingredients used in 211 products are registered for use in agriculture, forestry, mosquito/blackfly control, and homeowner situations (Table 1). The Office of Pesticide Programs (OPP) formally recognized in 1979 that microbial pesticides are distinct from conventional chemical pesticides and made the commitment to develop appropriate testing guidelines for microbial pesticides. The guidelines for microbial and biochemical pesticides were published in 1982 as the Pesticide Assessment Guidelines, Subdivision M. (The microbial section of Subdivision M was updated and revised in July of 1989.) In 1984, data requirements for microbial pesticides were codified in 40 CFR Part 158.170.

A. TOXIC SUBSTANCES CONTROL ACT (TSCA)

"Precursors of pesticides can in some instances be pesticide intermediates and subject to TSCA. For most pesticides the issue of whether some portion of the production process would trigger TSCA requirements is not relevant since the pesticide will be produced in such a way that there will be no pesticide intermediate. The most likely circumstance where triggering of TSCA requirements could arise involves the production scale fermentation of an intergeneric microorganism which is subsequently going to be processed (e.g., by lysis or fixation of the cells or purification of a substance) to produce the actual pesticide active ingredient."[1]

Table 1   **EPA currently registered microbial pesticides — December 10, 1992**

Microbial active ingredient	Year registered	Number registered	Pest/disease controlled
Bacteria			
Bacillus popilliae + B. lentimorbus	1948	5	Japanese beetle larvae
B. thuringiensis subsp. kurstaki	1961[a]	136	Lepidopteren larvae
Agrobacterium radiobacter	1979	2	Crown gall disease
B. thuringiensis subsp. israelensis	1981	27	Dipteran larvae
B. thuringiensis subsp. san diego[b]	1988	1	Coleopteran larvae
B. thuringiensis subsp. tenebrionis	1988	6	Coleopteran larvae
Pseudomonas fluorescens	1988	5	Pythium, Rhizoctonia, and frost inhibition by competition
B. thuringiensis subsp. kurstaki strain EG 2348	1989	1	Dipteran larvae
B. thuringiensis subsp. kurstaki strain EG 2424	1989	1	Dipteran/ Coleopteran larvae
B. thuringiensis subsp. kurstaki strain EG 2371	1990	2	Lipidopteran larvae
B. sphaericus	1991	1	Dipteran larvae
B. subtilis GBO3	1992	2	Damping off disease
B. thuringiensis subsp. aizawai strain GC-91	1992	2	Lepidopteran larvae
B. thuringiensis subsp. aizawai	1992	2	Lepidopteran larvae
P. syringae[c]	1992	2	Frost inhibition by competition
Fungi			
Phytophthora citrophthora	1981	1	Citrus stangler vine
Colletotrichum gloeosporioides	1982	1	Northern joint vetch
Trichoderma harzianum (ATCC 20476)[d]	1989	1	Tree wound decay
T. polysporum (ATCC 20475)[d]	1989	1	Wood rot
Gliocladium virens G-21	1990	2	Phthium, Rhizoctonia
T. harzianum Rifai strain KRL-AG2	1990	2	Damping-off disease
Lagenidium giganteum	1991	3	Mosquito larvae
Protozoa			
Nosema locustae	1980	6	Grasshoppers
Viruses			
Polyhedral inclusion bodies of Heliothis nucleopolyhedrosis virus (NPV)	1975	1	Cotton bollworm, bud worm
Polyhedral inclusion bodies of Douglas fir tussock moth NPV	1976	1	Douglas fir tussock moth larvae
Polyhedral inclusion bodies of gypsy moth NPV	1978	2	Gypsy moth larvae
Polyhedral inclusion bodies of Pine sawfly NPV	1983	1	Pine sawfly larvae

[a] *Bacillus thuringiensis* subsp. *thuringiensis* was registered in 1961 and was replaced by subsp. *kurstaki* in 1970; [b] The taxonomic designation for *B. thuringiensis* subsp. san diego has been changed to *B. thuringiensis* subsp. *tenebrionis* (San Diego isolate); [c] One registered product is sold in combination with *Pseudomonas fluorescens;* [d] *Trichoderma harzianum* (ATCC 20476) and *T. polysporum* (ATCC 20475) are always sold and used in combination.

B. FIFRA AND FFDCA REQUIREMENTS FOR MICROBIAL PESTICIDES

The main aspects of FIFRA as they relate to the pesticide registration process are summarized below. Any pesticide product that is to be used on a food crop to be distributed in commerce must have a tolerance (maximum legal residue level) or temporary tolerance under Section 408 or 409 of the FFDCA or must be granted an exemption from the requirement of a tolerance to avoid rendering food adulterated under the FFDCA. All microbial pesticides registered to date for use on food crops have been exempted from the requirement of a tolerance.

In the process of pesticide development, field testing is often necessary to evaluate the efficacy of a microbial pesticide. Title 40 CFR Part 172 describes when it is necessary to obtain an Experimental Use Permit (EUP) under Section 5 of FIFRA for testing unregistered microbial pesticides. Briefly, if the pesticide is (1) to be tested on a crop to be used as food or feed or on animals to be used as food or feed or (2) the size of the outdoor test acreage is greater than 10 acres of land or 1 surface acre of water, an EUP is required. The 10 and 1 acreage limitations are applicable only for outdoor terrestrial and aquatic uses. For those pesticides being tested on sites for which acreage is not relevant, the determination of the need for an EUP is made on a case-by-case basis. Consultation with the EPA is recommended when persons intending to conduct such tests are uncertain as to whether the testing may be allowed without a permit. For nonfood uses, it is generally presumed that an EUP is not required for outdoor field tests under 10 acres of land or 1 surface acre of water. Other criteria to determine when an EUP must be obtained are set forth in 40 CFR Part 172.3. An EUP is of limited duration and requires that the test be conducted under controlled conditions.

In 1984, the EPA recognized that there may be potentially significant impacts from the use of genetically altered and nonindigenous microbial pesticides in the environment, even at the small-scale testing stage. To address this concern, the EPA issued an interim policy statement (49 FR 40659) in 1984 that announced the presumption in the EUP regulation (40 CFR 172.3) that an EUP would not be required for small-scale testing and would not be applicable to such tests involving genetically altered and nonindigenous microbial pesticides. This policy requires notification of EPA prior to small-scale field testing of genetically altered and nonindigenous microbial pesticides so that the agency can determine whether an EUP is required. The 1984 interim policy was subsequently incorporated into a government-wide 1986 policy statement that retains the notification requirement (51 FR 23302). The Agency is currently working on revising the existing EUP regulation to codify the notification requirement.

In order to market a pesticide product, a company generally must obtain a Section 3 registration; however, there are two additional means under FIFRA whereby a pesticide product may be distributed in the absence of a registration or an EUP. The first of these is pursuant to an emergency exemption under Section 18 of FIFRA. Under this section, federal or state agencies may request limited approval for an unregistered use of a currently registered pesticide product or the use of an unregistered pesticide product. Such a request can only be granted when there is a potentially severe economic or human health impact and no other alternatives are available for pest control. A Section 18 exemption usually allows use of the particular pesticide product for a year; however, the duration of the exemption may be limited or expanded depending on the situation. In addition to emergency exemptions under Section 18, cases exist where a particular pesticide product may be registered for one or more uses, but not for a particular use which is determined by the State as being a special local need. In these cases, the State may register that use or formulation needed for the special local need under Section 24(c) of FIFRA provided that appropriate tolerances or exemptions from tolerance exist if food or feed uses are involved. The EPA has 90 days to disapprove of such State registrations. If the Agency does not respond, then that use and or formulation, previously not part of a federal registration, becomes part of either an existing or a new federal registration. (Refer to Section 24(c)(2) of FIFRA for specific details.)

1. Data Requirements

The applicant is required to submit supporting data as outlined in 40 CFR Part 158. The recommended test methods provided in the 1989 revised Pesticide Assessment Guidelines Subdivision M and corresponding data requirements in 40 CFR Part 158 are set forth in four basic areas: product and residue analysis, environmental fate, nontarget organism testing, and human health effects. The testing schemes for human health and nontarget organism effects are tiered, i.e., certain testing is not required unless triggered by results of initial testing. Residue analysis and environmental fate requirements are usually triggered by human health effects data and nontarget organism data, respectively. Following is a description of the data necessary to support a Section 3 registration.

2. Nontarget Organism Testing

The unique nature of microbial pesticides has led to changes in the data requirements for these products as compared to synthetic chemical pesticides. This is particularly evident in the area of assessing risk to nontarget wildlife. The testing requirements have been set up to test not only toxicity but also pathogenicity. This was accomplished by increasing the length of the tests (up to 30 days), and looking for signs of infection during and after the testing period. Beneficial insect testing was added in order to ensure that the risk from insect pathogens had been adequately assessed.

The specific nontarget organism data requirements used to assess the risk of biological control agents of postharvest diseases would also follow this case-by-case procedure. For example, in many instances the use of these products would be in enclosed areas (i.e., packinghouses, storage buildings, etc.) and would be considered an indoor use. If this were the case, then testing of nontarget organisms would probably not be required because of a lack of exposure. However, if the proposed use was determined to be outdoor and to have potential exposure to nontarget organisms, then the ecological testing requirements would need to be addressed.

At the first tier, short-term testing utilizes maximum hazard dosing of nontarget organisms. If no adverse results are observed in Tier I, then further testing is not warranted and environmental fate data are not required. If adverse effects are observed in the first tier, then potential exposure to nontarget organisms is evaluated in Tier II. In the first tier of nontarget organism testing, avian oral, freshwater fish, freshwater aquatic invertebrate, and honeybee testing are required. In addition, tests to evaluate microbial pesticide effects on wild mammals, plants, and beneficial insects are required depending on the proposed use site, target organism, and degree of anticipated exposure. It is important for applicants to work closely with the Agency at this point to ensure that the proper tests are done and any unique characteristics of the microbial pesticide are taken into account in specific testing procedures.

The current regulations allow for flexibility regarding required data. This is accomplished through the use of data waivers for requirements which are not appropriate for a particular product or for which there is enough information from the literature or other research data to adequately assess the risk of that microbial pesticide on a particular nontarget organism. In addition, the agency has the authority to invoke additional testing requirements if a potential risk to nontarget organisms has been identified and needs to be investigated. This flexible approach ensures that the potential risk presented by microbial pesticides to nontarget organisms will be properly assessed.

3. Mammalian Toxicology and Product Analysis
Data Requirements

The purpose of reviewing mammalian toxicology data for microbial pesticides is to ensure that the use of these products causes no unreasonable adverse effects to human health. In order to do this the agency must verify that the microbial product is correctly identified, is manufactured in a manner to prevent contamination with human pathogens, and presents little possibility of pathogenicity or toxicity to humans or other mammals. The focus of this section is to address the requirements of product identity, manufacture, and mammalian toxicology testing.

Crucial to any evaluation of the hazards presented by a microbial pest control agent is correct identification. This identification allows the Agency to ascertain possible hazards associated with closely related organisms and to utilize published literature to facilitate the review. The Agency expects an applicant to provide the most accurate, current taxonomic information to verify the identity of their active microbial agent. For bacteria this includes morphology, biochemical tests, and antibiotic sensitivity. Information for other types of microbes such as fungi, viruses, and protozoa is usually less extensive and may therefore involve serotyping, deoxyribonucleic acid (DNA) homology, RFLP, or isozyme analysis when available. Any adverse effects known to be associated with the microbe, or closely related species, should also be reported such as toxin production and pathogenicity in species other than the target pest. For genetically engineered microbes similar information is required for both the final microbe and the source of the inserted DNA as well as specifics on the DNA construct used.

The method used to manufacture microbial products is examined to determine whether adequate quality controls are in place to ensure a pure product. This review includes an examination to ensure that a method to verify purity and stability of the seed or stock cultures. Consideration is given to final quality control measures that determine potency to ensure that these tests relate to bioactivity. The final product should be examined for the presence of human bacterial pathogens such as *Vibrio, Shigella,* and *Salmonella* or provide evidence that these organisms are unable to thrive in the growth medium used. Producers of *Bacillus thuringiensis* products are in addition required to perform a subcutaneous injection test on mice for each technical batch to ensure that no *Bacillus anthracis* or other similar contaminants are present. The emphasis of the review is to ascertain that contamination by mammalian pathogens is improbable in the chosen production method. Inert ingredients within the actual end-use and manufacturing-use products are also reviewed to ensure that they have been used in other pesticide products or that they have sufficient data to support their use. If the pesticide products have food or feed uses, a verification that tolerance requirements have been met for all inert ingredients is done.

To assure the safety of microbial pest control agents toward mammalian species, the agency has adopted a tiered testing scheme. Tier I is designed to expose the test animal, mice or rats, to a single acute, maximum hazard dose of the live microbial pesticide. The maximum hazard dose is intended to provide a worst case exposure to the microbial pesticide to assure that the expected use rate will not result in unreasonable adverse effects toward humans. The dose is administered by the routes most likely to occur in product use: oral and pulmonary. A test by intravenous injection (or intraperitoneal injection for larger microbes such as fungi or protozoa) is done to permit a mammalian pathogenicity evaluation when the normal barriers to infection have been bypassed. Tests involving mammalian tissue cultures are required for viral pest control agents to ensure there is no possibility of mammalian infection given optimal conditions for expression of viral pathogenesis.

In addition to testing the safety of the actual microbial pest control agent, the safety of the whole/total pesticide product including inert ingredients is tested. Acute dermal toxicity, oral toxicity, inhalation toxicity, eye irritation, and dermal irritation testing may be required. However, waivers of these studies may be appropriate depending on the nature of the inert ingredients and results of the initial toxicity/pathogenicity tests with the microbial agent.

Any incidents of hypersensitivity in production workers, applicators, or the general public must be reported to the Agency.

The major end points for the toxicity/pathogenicity tests are to observe any adverse effects on the test animals and to establish that the microbial test substance is being cleared from the exposed animals. The animals are observed for any unusual clinical signs during the test, and for gross abnormalities at necropsy. Specific organs are isolated from sacrificed animals during the course of the test to determine the level of microbial test substance present. This is done to assure that a high dose was administered and track the normal mammalian response which recognizes the test substance as foreign and clears it from the system. Unusual persistence of the test microbe in an organ is also considered an adverse effect. Replication of the test microbe in organs is also an adverse reaction, indicating potential for infectivity.

If any adverse effects are noted in Tier I of the toxicity/pathogenicity tests, further testing is indicated using a tier progression to verify the observed effects and clarify the source of the effects. These Tier II tests could involve a subchronic toxicity/pathogenicity test or, if the adverse effect was believed to be due to a toxic reaction rather than pathogenicity, an acute toxicity test to establish an LD_{50} value for the toxin. Residue data are required if significant human health concerns arise from the toxicology testing. The majority of products screened to date have not indicated any adverse effects to warrant testing further than Tier I.

III. THE U.S. DEPARTMENT OF AGRICULTURE APHIS BIOLOGICAL CONTROL POLICY*

The Animal and Plant Health Inspection Service (APHIS) is currently reviewing its biological control responsibilities and procedures to clarify and renew its biological control policy. APHIS has statutory responsibilities for biological control (including regulation of importation, interstate movement, or release of biological control agents) and programmatic responsibilities (including implementation of biological control programs).

The two underlying principles of development of an APHIS biological control policy are that (1) when used appropriately, modern biological control is a desirable, cost-effective, and environmentally beneficial pest control strategy for protecting American agriculture and the environment which should be more fully utilized; and (2) APHIS can and should help increase utilization of biological control through appropriate regulation and procedures. This includes, in the context of this section, clarifying and improving procedures for importation, interstate movement, or release of biological control agents. There are six steps in this process:

1. Examine the legislative authority for biological control regulation in APHIS.
2. Develop an APHIS biological control philosophy.
3. Confirm lines of authority for biological control in APHIS.
4. Examine the current APHIS procedures for approval of biological control agents.
5. Revise regulations and procedures.
6. Begin a process of periodic review of the APHIS biological control regulations and procedures.

A brief discussion in developing the APHIS biological control policy follows.

---

* The views expressed in this article relative to APHIS are those of the authors and do not necessarily represent those of the United States government.

A. THE LEGISLATIVE AUTHORITY FOR BIOLOGICAL CONTROL REGULATION IN APHIS

APHIS is charged with the enormous responsibility of "Protecting American Agriculture". The regulations under which biological control is regulated are summarized in Shantharam and Foudlin.[2] These authors emphasized and stated, "APHIS has broad regulatory authority under the Plant Quarantine Act (PQA, 1912) and the Federal Plant Pest Act (FPPA, 1957)" and "APHIS is also committed to maintaining flexibility to accommodate the changes in the rapidly developing field of biotechnology," but the same is true for biological control.

In particular, APHIS has responsibility under the PQA and the FPPA for oversight of importation, interstate movement, or release of plant pests. For biological control, "plant pests" are defined broadly and include organisms which can have either a direct or indirect effect on plants. This definition currently includes beneficial natural enemies of target pest species. These are the "biological control agents" and include arthropods such as parasitic wasps and other parasitoids, predators, phytophagous arthropods, phytopathogens, antagonists, and competitors. The biological control agents can be either exotic or native to the United States, and are regulated without regard to the type of biological control program (classical/inoculative, augmentative/inundative, or by conservation of natural enemies). APHIS is also responsible for biological control of veterinary pests, but little has been done in this area to date.

B. THE APHIS BIOLOGICAL CONTROL PHILOSOPHY

APHIS uses its regulatory responsibility for biological control to facilitate the use of safe biological control agents. A critical document illustrating this commitment is the APHIS Biological Control Philosophy, which was approved by the Administrator, Mr. Robert Melland, on August 7, 1992. This philosophy* states:

> "APHIS believes that modern biological control, appropriately applied and monitored, is an environmentally safe and desirable form of long-term management of pest species. It is neither a panacea nor a solution for all pest problems. APHIS believes that biological control is preferable when applicable; however, we also recognize that biological control has limited application to emergency eradication programs. Whenever possible, biological control should replace chemical control as the base strategy for integrated pest management.
>
> In support of this philosophy, APHIS will develop regulations that facilitate the release of safe biological control agents, while maintaining adequate protection for American agriculture and the environment. The regulations will give clear and appropriate guidance to permit applicants, including specific types of data needed for review and environmental analysis and specific time limits for Agency review. They will be updated as science progresses. APHIS believes that public input on procedures to approve the release of biological control agents is a desirable and necessary step, and will strive to gather input from scientists, industry and the public."

C. LINES OF AUTHORITY FOR BIOLOGICAL CONTROL IN APHIS

The lines of biological control authority in APHIS have also recently been examined and clarified. The key groups in biological control in APHIS are Plant Protection and Quarantine (PPQ), Biotechnology, Biologics, and Environmental Protection (BBEP), and the National Biological Control Institute (NBCI).

---

* Further details about biological control in APHIS, including a copy of the APHIS Biological Control Philosophy, can be obtained from the National Biological Control Institute, USDA, APHIS, OA, Federal Building, Room 538, 6505 Belcrest Road, Hyattsville, MD 20782.

The responsibility for authorizing the use and introduction of biological control agents rests within PPQ under the authority of the Federal Plant Pest Act. The Plant Pest Permit Section (PPPS) of the Biological Control and Taxonomic Support group of PPQ is responsible for the permitting process for biological control agents in APHIS. Other groups function to support the decisions of PPQ. For example, environmental analysis and documentation is provided by BBEP. The NBCI provides policy advice to the APHIS administrator, facilitates development and review of biological control regulations, and supports other areas of biological control, such as providing technical support for PPQ Biological Control Operations programs. International Services helps coordinate classical biological control and biologically based control (such as use of the sterile technique) in foreign countries. It is possible that two other APHIS groups (Veterinary Services and Animal Damage Control) will become involved in biological control in the future.

D. CURRENT APHIS BIOLOGICAL CONTROL PROCEDURES

NBCI has coordinated documentation by PPQ and BBEP of the current procedures used by APHIS to facilitate use of safe biological control agents. This is the first step in revising the APHIS biological control regulations. A detailed assessment was made of the current APHIS approval procedures which serves as a springboard for improving the system by pointing out its complexity and the opportunities for streamlining it, without jeopardizing the safe use of agents.

Applicants for APHIS permits (for importation from overseas to quarantine in the United States, interstate movement, or field release) should contact Deborah Knott, Head PPPS in Matt Royer's program (see following page for contacts). Mrs. Knott will advise applicants on procedures, notify EAD, and otherwise facilitate the process.

E. REVISE REGULATIONS AND PROCEDURES

A draft revision of the APHIS biological control regulations is also being prepared. NBCI has facilitated this review by seeking substantive, constructive input on the adequacy of the current procedures from APHIS and users. APHIS will be publishing a notice of proposed rule making in the *Federal Register* and accepts and welcomes public comment regarding the proposed rule. The new regulations will:

- Facilitate the process of obtaining permits for importation, interstate movement, or release of safe biological control agents
- Document the pest management options and the reasons for decisions about permit issuance or denial
- Emphasize a flexible, scientific process (use of data bases, place decisions in a risk assessment context, etc.)

The head of the PPPS of PPQ will continue to be the main point of contact in APHIS for the permitting process for biological control. Similarly, draft procedures (a "User's Guide") are being prepared in parallel with the regulation, which will be distributed for comment by scientists and others.

F. BEGIN A PROCESS OF PERIODIC REVIEW OF THE APHIS BIOLOGICAL CONTROL PROCEDURES

It is planned that the APHIS biological control procedures will be tested for a reasonable period (e.g., 2 to 3 years). Comment will continually be sought during this period from regulators, scientists, environmental groups, and others. The procedures may again be modified as a result of this input. This process will be repeated every few years.

IV. CONCLUSIONS

For many years the APHIS, the EPA, and other federal and state agencies have been in communication on the oversight of biological control agents. This dialogue will continue. Additionally, closer links with other federal, state, and international organizations involved with biological control are being established by the NBCI. It is hoped that this process will ensure meaningful contact between regulators, reduce confusion by applicants, and help keep biological control agents at pace with the science of biological control and the needs of American agriculture, industry, and the environment.

CONTACTS

EPA CONTACTS
Microbial Insecticides
Phillip Hutton
Product Manager 18
(703) 305-7690

Microbial Herbicides and Fungicides
Sidney Jackson
Acting Product Manager 21
(703) 305-6900

Plant Growth Regulators
Cynthia Giles-Parker
Product Manager 22
(703) 305-5540

TSCA
David Giamporcaro
Section Chief, Biotechnology Program
OPPT/CCD/PDB
(202) 260-6362

USDA, APHIS BIOLOGICAL CONTROL CONTACTS
Biological Control Policy and General Information
Ernest Delfosse
Director, National Biological Control Institute
Telephone (301) 436-4329   Fascimile (301) 436-7823

Biological Control Regulation and Permits
Deborah Knott
Head, Plant Pest Section
Plant Protection and Quarantine
Telephone (301) 436-505   Fascimile (301) 436-8700

Environmental Assessments
Carl Bausch
Deputy Director, Environmental Analysis and Documentation
Biotechnology, Biologics, and Environmental Protection
Telephone (301) 436-8565   Fascimile (301) 436-8669

ACKNOWLEDGMENTS

The authors gratefully acknowledge the assistance of the following individuals who helped with gathering information and editorial review of this manuscript: U.S. EPA — Willie Nelson, Patricia Cimino, Linda Hollis, Laura Dye, Amy Rispin, Sharlene Matten, William Schneider, Melissa Chun, Fred Betz, Pat Roberts, Jon Fleuchaus, Jim Beech, Clayton Beegle, and David Giamporcaro; USDA/APHIS — Lonnie King, Sally McCammon, Norm Leppla, Carl Bausch, Bob Flanders, Michael Oraze, and Debbie Knott.

REFERENCES

1. **Matten, S., Schneider, W., Slutsky, B., and Milewski, E.,** Biological pesticides and the U.S. Environmental Protection Agency, in *Advanced Engineered Pesticides,* Kim, L., Ed., Marcel Dekker, New York, 1993.
2. **Shantharam, S. and Foudlin, A.,** Federal regulation of biotechnology: jurisdiction of the U.S. Department of Agriculture, in *Biotechnology for Biological Control of Pests and Vectors,* Maramorosch, K., Ed. CRC Press, Boca Raton, FL, 1991, 239.

FURTHER READING

Betz, F., Forsyth, S., and Stewart, W., Registration requirements and safety considerations for microbial pest control agents in North America in Marshall, L., Lacey, L., and Davidson, E., Eds. *Safety of Microbial Insecticides,* CRC Press, Boca Raton, FL, 1990, 239.

Mendelsohn, M., Rispin, A., and Hutton, P., Environmental Protection Agency oversight of microbial pesticides, in *Dispersal of Living Organisms into Aquatic Ecosystems,* Rosenfield, A. and Mann, R., Eds., Maryland Sea Grant College, College Park, MD, 1992.

Microbial Pesticides; Interim Policy on Small-Scale Field Testing, FR 49 (202), 1984, 40659.

Coordinated Framework for Regulation of Biotechnology; Announcement of Policy and Notice for Public Comment, FR 51 (123), 1986, 23302.

U.S. Congress, Federal Insecticide, Fungicide, and Rodenticide Act (FIFRA) of 1972, 7 USC 136 *et seq.,* as amended October 24, 1988.

U.S. Congress, Federal Food, Drug, and Cosmetic Act (FDCA), and as amended, 21 USC 201 *et. seq.,* 1991.

U.S. Congress, Toxic Substances Control Act (TSCA), and as amended. 15 USC 2601 *et seq.,* 1976.

Code of Federal Regulations (CFR), Title 40, Washington, D.C., 1991, Parts 152.20 158, 172, and 180.

Chapter 12

Natural Fungicides and Their Delivery Systems as Alternatives to Synthetics

Horace G. Cutler and Robert A. Hill

CONTENTS

I. INTRODUCTION

In *The Ascent of Man,* Jacob Bronowski[1] relates the problem that Leo Szilard was having relative to the concept of the chain reaction in the context of atomic energy. In 1933 Szilard was living at the Strand Palace Hotel in London, one of his favorite haunts, and working at Bart's Hospital. On his way to work one morning, Szilard had nothing in particular on his mind (according to Bronowski's verbal account in the British Broadcasting Company's production of "The Ascent of Man"). Threading his way through the crowds of pedestrians, he came to an intersection just as the light was changing from green, to amber, to red. By the time the light had changed to green Szilard understood the nature of the chain reaction. That is, a single neutron hits an atom, the atom breaks into two parts, and a chain reaction is created. He filed a patent in 1934 in which the words "chain reaction" are prominent. These events are of interest for two reasons. First, Szilard was having difficulty formulating a specific aspect in atomic energy relative to a certain sequence; and second, while the problem was lodged in his memory, a mundane event like the sequencing of a traffic light, provided the impetus needed for a successful resolution of a physical reaction both spatially and mathematically. The remarkable point of this story is that problems may be sequestered in the memory along with odd bits and pieces of apparently useless information and at a time when one is not concentrating on the solution, a crystallization takes effect and rudimentary answers emerge which can then be developed to deliver practical solutions. It may be that many of the solutions to the problem of replacing synthetic fungicides are already at hand, though scattered and fragmented, and have yet to be pieced together.

There are several questions surrounding the application of fungicides on crops. Among these are, "Will the presently used fungicides, which are purely synthetic, be removed from the market within, for example, the next 5 years?" The answer may well be yes. Some have been questioned because of their safety and others will cost too much to reregister. Another question is, "Can substitutes be found and, if so, how may they be applied for maximum benefit?" Again, there are a number of natural products which may be specifically targeted as antimicrobials, which are nonpersistent and which may be delivered by unique systems to effect a maximum response. In addition, the delivery system may be the microorganism that produces the antimicrobial secondary metabolite.

Statements about biocontrol agents, their biologically active natural products, and the use of either the microorganism or the natural products or a combination of the two, require supporting evidence; and that we intend to do in this presentation. However, before examining that evidence, it is necessary to clarify any misunderstandings that may arise concerning natural products and synthetics. For the sake of this discussion, a natural product is a secondary metabolite of fungal or plant origin. A synthetic chemical is one that has been manufactured in a laboratory and whose template has not been constructed on a biorational basis. That is, the compound is the clever product of a chemist's mind, imagination, and skill. However, there does exist the situation where a natural product may be synthesized in the laboratory and it is identical in every physical way to the material obtained from nature. We consider the latter, even though it may be of "synthetic" origin, to be a natural product.

II. PRODUCER ORGANISMS AS DELIVERY SYSTEMS

During the course of isolating biologically active natural products from fungi, applying them to plants, and using the producer organism as biocontrol agents, we have been struck by a number of points. Among these are the observations that microorganisms grown *in vitro* produce the sought metabolite within a fairly narrow window of time, maybe as little as 2 to 3 days, and is then seen no more. However, in the natural state, where nutrients are abundant, they produce secondary metabolites in a continuum which appears to last some weeks or even longer. This is the case in certain biocontrol situations. The other striking point is that they readily secrete those metabolites into their immediate surroundings with relative facility; and in addition they may deliver the materials to critical active sites and tissue conduits. Conversely, a biologically active natural product may be isolated from a producer fungus which has been used as a biocontrol agent and applied to a crop as a bolus. The bolus may decay at a predictable rate depending on temperature, light, and degradation by microorganisms; and some of the material will be lost at time of application by aerosol while some may not penetrate to the active sites. Simply stated, the ideal delivery system seems to be the microorganism which, in a competitive environment, produces the desired metabolite over a period of time. A classical example, while it is not a fungicidal one, serves to illustrate the point; and while the history is now over 60 years old and has been available in English for over 40, the lessons that it has to offer are valuable and may have been overlooked.

In 1926, a paper was presented by Eiichi Kurosawa titled, "Experimental Studies on the Secretion of 'Bakanae' Fungus on Rice Plants," in which it was noted that rice plants infected with the Bakanae fungus grew unusually tall;[2] and it was thought that the hyperelongation was induced by some chemical action. Solid or liquid cultures of the organism gave rise to sterile culture filtrates which, when fed to rice and grass seedlings, caused the leaves to elongate. In 1938, Yabuta and Sumiki[68] reported the isolation of gibberellin A and B from the Bakanae fungus, now called *Gibberella fujikuroi*, in pure crystalline form; and by 1950 a number of publications concerning gibberellic acid had appeared. Today, the commercial use of gibberellic acid is mostly limited to GA_3 and the compound has a number of horticultural applications as a plant growth regulator. Uses on crops range from blueberries, cherries, grapes, citrus, rhubarb, asparagus, beans, celery, cucumbers, lettuce (seed), parsley, peas, seed potatoes, barley (malting), hops, strawberries, spinach, and Bermuda golf turf to ornamentals and sugarcane (in Hawaii).[3] The application range is from ~10 mg/l in hops to 2000 mg in 7 gal of water per acre for sugarcane. It is also used in rice to promote germination and growth of semidwarf varieties and is sold under the trade name of Release™. Treated seed can be sown deeper than normal; and seedlings, which are taller as a result of treatment, compete well against weeds. If one considers a rice treatment to be in the order of 100 mg/l, then in order to obtain that amount from fermentation — assuming production by liquid fermentation — a maximum yield would be in the order of 100 mg/l under ideal conditions — from experience with fermentation systems. Thus a relatively considerable biomass is needed *in vitro* to produce these amounts to induce a response in a given crop when the material is applied as a bolus vs. the amount produced *in vivo* in an infected seedling. In this regard it is important to reflect upon Sawada's statement made in 1912 concerning the effect of *Gibberella fujikuroi* on rice plants. He says, "On microscopic examination the plant system is found to contain mycelium. It is thought that the plants grow taller due to some stimulation from the mycelium."[4] Certainly, the fungal infection must have been difficult to see with the naked eye and required microscopic examination. Additionally, the amount of fungal biomass must have been small compared to the plant biomass; and consequently the organism was an efficient delivery system for gibberellic acid. Unfortunately, *Gibberella fujikuroi* is a phytopathogen and to date has no practical use as a biocontrol organism for along with GA_3 occurs the plant growth inhibitor 5-*n*-butylpicolinic acid; and it has been suggested that fusaric acid may also be a phytotoxic metabolite that results in seedling death.[4] There is the possibility that a strain of *Gibberella fujikuroi* exists that produces GA_3 but not the phytotoxins, and that strain would be a useful biocontrol organism.

III. BIOCONTROL AND ORGANIC CROP PRODUCTION

To those involved with the industrial research and development of synthetic fungicides, especially during the 1950s and 1960s, the concept of organic crop production, even on a small scale in home vegetable gardens, appeared to be altruistic and impossible. However, disease suppression using composts and mulches was described by Zarathustra 8000 years ago in his *Zend Avesta;* and it was in this document that an holistic approach to crop production was first recorded. These approaches have since been used by organic gardeners.[5] While the natural suppression of plant diseases has been recognized, but imperfectly understood, for at least a century,[6] the deliberate use of biological control agents for disease control is recent relative to that of biological control of insects and weeds.[7] It is especially important to note that many of the soils that suppress plant diseases are particularly high in organic matter and that it is the organic matter that supports the growth of the beneficial microorganisms.[8-10] These include such species as *Trichoderma* and *Pseudomonas* which are known (*vide infra*) to suppress the activity of soilborne plant pathogens. Additionally, two types of disease control have been identified, the first being a "short-term" suppression of pathogens brought about by increased microbial activity and the second being a "long-term" suppression that is influenced by a number of factors.[8]

IV. REPLACING SYNTHETIC FUNGICIDES IN HORTICULTURAL AND TIMBER CROPS

A particularly intensive effort is underway in New Zealand to replace synthetic fungicides in horticultural and timber crops with biocontrol agents and natural products. The reasons for this are largely economic because the country depends on its export market, especially to the Pacific Area, for revenue. Importing countries are now scrutinizing horticultural produce and timber for pesticide residues. Experiments have led from the application of biocontrol agents for controlling fungi to a chemical rationale for the mode of action of the agent.

A. *ARMILLARIA*: AN IMPORTANT PHYTOPATHOGEN IN NEW ZEALAND

Armillaria, a fungal pathogen of forest trees, was first identified 115 years ago; and it is now recognized as a major problem in a variety of woody plant species worldwide, so that more than 500 different plant species are known to be susceptible to the organism.[11] In undisturbed forests and native bush the organism rarely causes serious damage worldwide;[12,13] but when trees are harvested, the rotting stumps and roots provide a rich source of nutrients so that the *Armillaria* may become destructively infective to any remaining shrubs and trees. *Armillaria* can be devastating to the forest industry, and billions of dollars are lost annually because of lost timber. The worst losses follow reforestation after clearing the natural tree cover.[14-20] In New Zealand, losses of less than 5 to over 90% in *Pinus radiata*, a major timber crop for local and export markets, have been attributed to *Armillaria*.[21]

Another major crop affected by *Armillaria* is kiwifruit, and the orchards are planted on cleared lands. In order to protect the plantings, they are surrounded by tall windbreaks that generally consist of closely trimmed willow trees, sometimes *Cryptomeria, Casuarina,* and occasionally bamboo. The disease was listed as a new one for kiwifruit in 1955 in New Zealand. However, the first detailed account of *Armillaria* infection in kiwifruit was in a U.S. Department of Agriculture orchard in California; and it described the decline and death of the vines from 1967–1971.

Before 1980, the incidence of *Armillaria* in New Zealand kiwifruit was only occasional, and it was considered to be a minor phytopathogen. Between 1980 and 1990 a dramatic increase occurred in the number of infected orchards, and the industry suffered as a consequence. It has been projected that if the present rate of the disease continues, then by 1995 over 2000 orchards will become infected at an annual loss of NZ $20 million (NZ $1 = U.S. $0.55); and three scenarios for the incidence of infection and economic loss are shown in Table 1.[22]

Both kiwifruit and *Pinus radiata* are major export crops for New Zealand, and treatment of these commodities with synthetic pesticides poses problems. First, importers in the Pacific are now scrutinizing products for pesticide residues; and second, New Zealanders are determined to protect their environment. Hence, biocontrol agents and biodegradable biologically active natural product pesticides are acceptable alternatives in agriculture. Insofar as controlling *Armillaria* is concerned, in 1936 Leach[12] reported some benefit from ring-barking forest trees; however, this is not advisable in New Zealand because willow trees treated this way appear to have high incidence of *Armillaria*, and willow is one of the shelter trees used in kiwi orchards.

B. *TRICHODERMA* AS BIOLOGICAL CONTROL AGENT

Trichoderma species have been used successfully in field trials to control many crop pathogens. Examples include *Nectria galligena* in apples,[23] *Sclerotium rolfsii* in tobacco, bean, iris;[24-27] *Rhizoctonia solani* in radishes, strawberries, cucumbers, potatoes, and tomatoes;[24,28-30] *Sclerotium cepivorum* in onions;[31] *Macrophomina phaseolina* in maize, melons, beans, and other economically important crops;[32] *Fusarium oxysporum* in tomatoes

Table 1 Incidence of *Armillaria* infections[a] in New Zealand kiwifruit
orchards and estimated cost ($NZ)[b] of the disease to the industry

Year	1980	1985	1990	1995[c]	1995[d]	1995[e]
No. of infected orchards	10	43	250	2353	1470	588
Losses[f] from *Armillaria*	0.09	0.37	2.13	20.0	2.5	5.0

[a] Number of orchards infected (1980–1990) based on figures for confirmed *Armillaria* infections; an underestimate of total; [b] Cost estimate based on average cost of disease isolation and yield loss for a representative Bay of Plenty (North Island) kiwifruit orchard; [c] Extrapolation from present trend; [d] "Moderate" grower uptake of prevention and control measures and research success "fairly good"; [e] "Excellent" grower uptake of prevention and control measures and research success "very good"; [f] $ million.

and *Chrysanthemum*[33,34] and *Verticillium albo-atrum* in tomatoes;[35] *Chondrostereum purpureum* in stone fruits and other crops;[36-38] and *Botrytis cinerea* in apples.[39] Papavizas[40] has comprehensively reviewed the potential of *Trichoderma* as a biocontrol agent.

C. THE SEARCH FOR BIOLOGICAL CONTROL AGENTS OF *ARMILLARIA* IN NEW ZEALAND

The most effective biological control agents for *Armillaria* in New Zealand include isolates of *Trichoderma hamatum* (Bon.) Bain, *T. harzianum* Rifai, *T. viride* Pers. ex S. F. Gray, and other *Trichoderma* spp. — particularly those collected from *Armillaria*-infected orchards and forest sites in the Bay of Plenty. Some *Trichoderma* strains were isolated from situations in which they were growing on and consuming *Armillaria* mycelium and rhizomorphs. On transfer to the laboratory, *in vitro* tests confirmed the activity of the *Trichoderma* isolates against *Armillaria;* and as the result of many tests, superior strains were selected for field use, and different fermentation and formulation technology is presently underway.

In collaborative research with the University of Auckland, which was initiated in 1988 to investigate the *in vitro* interactions between *Trichoderma* isolates and *Armillaria novae zelandiae* using dual plate techniques and visualization with a light and scanning electron microscope, 11 potentially superior isolates of *Trichoderma* were evaluated. These included strains of *T. hamatum, T. harzianum,* and *T. viride;* and the evaluations covered two major points: the antagonistic potential against *Armillaria* and the compatibility of the *Trichoderma* isolates with each other so that they could be used in an inoculum blend. All the *Trichoderma* isolates antagonized *Armillaria* in dual culture; and the antagonism was manifest by the formation of brown residues on the surface of the *Armillaria* mycelium, yellowing of the *Armillaria* mycelium, overgrowth of the *Armillaria* by *Trichoderma,* and extensive rhizomorph initiation of the *Armillaria* colony. Importantly, there were differences in the antagonistic response of the accessed *Trichoderma* isolates to *Armillaria;* and *in vitro* cultures of *T. harzianum* were easily overgrown by *T. hamatum* and *T. viride* in paired assays. Thus, the inclusion of isolates of *T. harzianum* in an inoculum blend with *T. hamatum* and *T. viride* would be undesirable.

In addition to the above observations, a temperature effect on the antagonism between *Trichoderma* and *Armillaria* was noted. The greatest antagonism was exhibited by *T. hamatum* and *T. viride* isolates between 20 and 25°C, while *T. harzianum* isolates were only effective at 25°C. There was also a pH effect on the antagonism between *Trichoderma* and *Armillaria;* and this was greatest at a basic pH on malt extract agar, while on tap water agar acidic conditions were generally more favorable. Furthermore, the germination of *Trichoderma* spores on a low nutrient medium was enhanced under acidic conditions. There was competition for nutrients between *Trichoderma* and *Armillaria* in dual culture due to differences in the relative growth rates. Interactions between *Trichoderma* and *Armillaria* rhizomorphs indirectly indicated that hyperparasitism may be part of the control mechanism.

D. ANTIBIOTIC PRODUCTION BY *TRICHODERMA*

Antibiotics were produced by some of the *Trichoderma* isolates *in vitro* in the New Zealand experiments, and antibiosis was detected using liquid culture and split plate techniques. However, the ability of the *Trichoderma* isolates to produce volatile and nonvolatile antibiotics was found to differ within and between species. The culture filtrates of some of the isolates were also found to be inhibitory toward the growth of *Armillaria*.[41]

E. MECHANISM OF BIOLOGICAL CONTROL BY *TRICHODERMA*

Recent research has shown that various *Trichoderma* species produce a number of antibiotics. The most common of these is 6-pentyl-α-pyrone (= 6-amyl-α-pyrone) (Figure 1)[42,43] which has potent antifungal activity. Its coconut/celery-like odor permeates the atmosphere on isolation, and can be easily detected in *Trichoderma* cultures by sniffing. *In vitro* assays with 6-pentyl-α-pyrone showed, for example, that a 1:40 dilution of the metabolite applied at the rate of 15 μl/4 mm disk inhibited the growth of *Aspergillus flavus*, the producer of aflatoxins.[42]

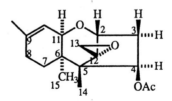

Figure 1   Structure of 6-pentyl-α-pyrone (= 6-amyl-α-pyrone).

Both *T. lignorum* and *T. viride* produce trichodermin (4β-acetoxy-12,13-epoxy trichothec-9-ene) — a natural product that has marked antibiotic effects against many fungi, including *Candida albicans* — but is relatively inactive against bacteria[44] (Figure 2). Unfortunately, it also possesses plant growth regulatory properties and is selectively toxic to certain herbaceous plants.[45] However, it has relatively low toxicity in mice (LD_{50} 1g/kg orally) compared to its congeners; and at one time was considered by the pharmaceutical trade to be a candidate antibiotic. A variety of other *Trichoderma* metabolites with biological activity have subsequently been discovered and are discussed later (*vide infra*).

Figure 2   Structure of trichodermin (4β-acetoxy-12,13-epoxytrichothec-9-ene).

F. EFFECTS OF *TRICHODERMA* AND 6-PENTYL-α-PYRONE ON *ARMILLARIA* IN *PINUS RADIATA*

Crude extracts from *Trichoderma* containing 6-pentyl-α-pyrone and synthetic 6-pentyl-α-pyrone, hereafter referred to as 6-amyl-α-pyrone to distinguish between the natural product and the synthetic copy of the natural product, were evaluated with *in vitro* assays against *Armillaria novae zelandiae*. Potent antimicrobial activity was seen with as little as 4 µl per disk with 6-amyl-α-pyrone, and concomitantly the crude extract was active. Other microorganisms were also strongly inhibited; and these included *Botrytis cinerea, Sclerotinia sclerotiorum, Chondrostereum purpureum, Phytophthora fragariae, Pythium ultimum,* and *Corticium rolfsii,* all important phytopathogens. These results led to field trials in their respective crops of importance.

As an initial step, because *Trichoderma* treatments appeared to be the most efficient delivery system for 6-pentyl-α-pyrone to the necessary sites, selected *Trichoderma* spp. isolates were tested in laboratory assays with *Pinus radiata* tissue cultured plantlets. No pathogenicity or toxicity was seen, except in very aged cultures where nutrients were exhausted. Following this, forest trials were initiated in January 1991 (summertime in New Zealand), and following treatment with *Trichoderma,* treated trees showed less mortality and were more vigorous compared to control treatments. Far fewer treated trees (5.9%) were infected and died from *Armillaria* compared with controls (22%) ($P < 0.019$). Treated trees were taller and had thicker trunks and wider canopy than untreated trees. Consequently, another 50 ha of *P. radiata* have been treated with *Trichoderma* and various combinations of *Trichoderma* and 6-amyl-α-pyrone to determine effects on *Armillaria* and enhancement of vigor.

G. EFFECTS OF *TRICHODERMA* AND 6-PENTYL-α-PYRONE/6-AMYL-α-PYRONE IN KIWIFRUIT

The stumps of shelter trees that had been cut down and were possible sources of *Armillaria* infection have been treated with *Trichoderma* formulations. Soil amendments have inhibited or prevented the spread of the organism within kiwifruit orchards, and in addition soil treatment in barrier trenches between infectious *Armillaria* sites and kiwi plantings have been very successful. Soil drenches, too, have been effective. Injections with formulations of *Trichoderma* directly into the trunks of kiwi vines have shown that infected plantings may recover; and pastes made up of *Trichoderma* applied directly to infected areas, where as much as four fifths of the vascular cambium has been destroyed, have completely healed the vines. As the vascular cambium grew, the vines regained their lost vigor and became productive. Root treatments with *Trichoderma* have reduced mortality in kiwifruit vine replants at diseases sites from approximately 50% of untreated plants to 5% of treated ones. Selected *Trichoderma* isolates have also been evaluated for antifungal use on stored kiwifruit, and *Botrytis cinerea* was totally inhibited. Other storage organisms, including *Sclerotinia sclerotiorum,* treated with species of *Trichoderma* and *Gliocladium* were successfully controlled for the first time in kiwifruit.

Armillaria infected kiwifruit vines in the Bay of Plenty were injected in February 1992 with treatments ranging from 10 to 100 µl per vine of 6-amyl-α-pyrone; 10 to 50 µl per vine of 6-pentyl-α-pyrone (the natural product is more difficult to obtain in quantity relative to the synthetic material); and 300 µl of a crude extract, known to contain 6-pentyl-α-pyrone, from a high yielding isolate of *T. hamatum*. Other infected vines were injected with mixed strain *Trichoderma* formulations with proven efficacy against *Armillaria*. All untreated *Armillaria* infected vines died within 6 months. Both 6-amyl and 6-pentyl-α-pyrone treatments significantly increased the survival rate (to ~50%) in infected vines. However, *Trichoderma* formulations were even more effective, and over 80% of the infected vines survived; while the crude extract was approximately as active as the 6-amyl and 6-pentyl-α-pyrone.

H. EFFECTS OF *TRICHODERMA* ON THE CONTROL OF SILVER-LEAF DISEASE

Effective disease control using high 6-pentyl-α-pyrone producing strains of *Trichoderma,* especially *T. hamatum,* has been achieved in the North Island of New Zealand against silver-leaf disease (*Chondrostereum*), an organism that was controlled *in vitro* by *Trichoderma* isolates in laboratory assays. Injections with liquid formulations of *Trichoderma* gave rapid control of silver-leaf in *Pyrus serotinia* (nashi, Asian pear) with even severely affected trees recovering completely. Most treated trees remained disease free 2 years following treatment. In addition, a pruning paste containing *Trichoderma* greatly reduced the spread of silver-leaf in infected nashi orchards. Research is now underway to investigate the efficacy of 6-amyl-α-pyrone against silver-leaf disease.

I. *TRICHODERMA* AND 6-PENTYL-α-PYRONE

Results from laboratory and field trials involving *Trichoderma* species as biological control agents indicate that the mechanisms of action of the most effective *Trichoderma* isolates involves 6-pentyl-α-pyrone and that the organism is an efficient delivery system for the active secondary metabolite. The synthetic natural product 6-amyl-α-pyrone has shown promising *in vitro* and *in vivo* control of several major phytopathogenic fungi affecting New Zealand crops, some of which have not been reported here. However, establishment and growth of 6-pentyl-α-pyrone producing *Trichoderma* biological control agents in living tissues enables production of the metabolite over an extended period of time and delivery to the site where activity is most effective.

V. REPLACING SYNTHETIC FUNGICIDES IN POSTHARVEST TREATMENTS

Horticultural produce may be treated with fungicides immediately following harvest to increase shelf life. This is a critical stage because the treatment may be persistent; and depending on the nature of the fungicide, the implications as far as the consumer is concerned may be of enduring consequence. Some biocontrol alternatives to synthetic fungicides have been evaluated and the chemistry studied in some detail.

A. *BACILLUS SUBTILIS* ANTIBIOTICS, *B. SUBTILIS* B-3, AND THE ITURINS

We have earlier discussed the fact that many microorganisms, though morphologically identical, produce different secondary metabolites. This can be enigmatic to the chemotaxonomist; but to the those who have diligently searched for new microbial natural products, it is no surprise to find that, indeed, many phenotypically identical microorganisms produce vastly different sorts of secondary metabolites each of which may be highly target specific.[46] *Bacillus subtilis*, its many strains, and diverse metabolites eloquently describe this dilemma. Depending on the source and presumably because of environmental pressures, the organism produces the cyclic peptides mycobacillin [47, 48] (Figure 3); subsporins[49] which consist — in subsporin A — of L-aspartic acid (four amino acid residues), D-aspartic acid,[2] L-glutamic acid,[2] D-tyrosine,[2] L-proline,[2] and ammonia.[5] Subsporin B lacks D-serine which is replaced by D-alanine. Subsporin C, as A, has no D-alanine but has four residues of D-aspartic acid; and L-glutamic acid has no ammonia.

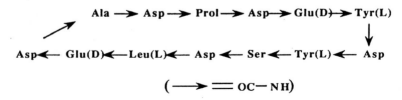

Figure 3   Structure of mycobacillin.

B. subtilis also produces mycosubtilin[50] which contains 45% aspartic acid, 5.4% tyrosine, 4.2% proline, and bacillomycin;[51] the latter contains glutamic acid, aspartic acid, tyrosine, serine, threonine, alanine, valine and isoleucine. Fungistatin[52] contains lysine, serine, aspartic acid, proline, threonine, alanine, tyrosine, tryptophan, valine, and isoleucine. The most important agronomically at present is iturin A[53] (Figure 4); and the other iturins, D and E[54] including the apparently biologically inactive iturin C,[55] are of interest because of their structural changes that alter activity. Bacilysin and fengimycin[56] are also *B. subtilis* cyclic peptides. Thus, we see that *B. subtilis* is very adept at utilizing amino acids in sequences that generate different antibiotics.

Even within the iturin-producing strains of *B. subtilis,* the congeners produced and consequently the efficacy to control certain phytopathogens may vary. For example, Phae et al.[57] collected *B. subtilis* from 53 well-matured compost heaps: 30 were from bark composts, 6 were from sewage, and 13 were from cattle manure. From these, four particularly potent *B. subtilis* strains were isolated; and of these two, NB-22 from night soil sludge and UB-24 from garbage (bark) were particularly active antimicrobials. Each contained a mixture of iturins, and the ratios of each were such that culture broths elicited different responses both *in vitro* and in greenhouse tests. For example, NB-22 produced a 95 to 100% inhibition in *Rhizoctonia solani, Pyricularia oryzae, Cochliobolus miyabeanus, Xanthomonas oryzae,* and *Pseudomonas lachrymans.* While NB-24 was equally as active against *R. solani, X. oryzae,* and *P. lachrymans,* it was only 80 to 90% active against *P. oryzae* and *C. miyabeanus.* The evidence for the difference between the various congeners of the iturins contained in the culture broths was shown by fast atom bombardment mass spectral data.[57]

There can be no doubt that faced with the subtle changes in iturin production within given strains of *B. subtilis* not only because of variability, which results from a change in the substrate on which the organism may be grown for commercial use, but also because of the source of accession, the biocontrol industry will have to initiate rigorous quality control for each batch of material produced. From the positive aspect, it may also be possible to use mixed batches of *B. subtilis,* each fermented separately and then custom mixed to be used under specific conditions to control a select group of microorganisms in a given crop. That is, the fungicide would have broader spectrum properties, but must be tailored to fit a field problem.

From another aspect, the iturins are a family of chemical structures that capture the imagination of medical and synthetic chemists because of their construction and potential to be slightly modified either by chemical synthesis or biotransformation, or a combination of the two. That biotransformation does occur during fermentation is obvious from our previous discussion. Of particular interest is iturin A (Figure 4) which, as its congeners, is a mixture of the "natural" L-amino acids and the "unnatural" D-amino acids in the following cyclic configuration: L-asparagine, D-tyrosine, D-asparagine, L-glutamine, L-proline, D-asparagine, L-serine, and a ß-amino acid.

It is curious that D-asparagine appears twice and L-asparagine once among the seven amino acid residues. This leads to the obvious question as to what happens if the sequences are changed or amino acids are substituted with other amino acids. Fortunately, nature has partially answered this question. Iturin C differs from iturin A in one amino acid residue; iturin A has an L-asparagine residue, but in iturin C that is replaced by an L-aspartic acid residue and as we have already stated, there is no antifungal activity.[54] Iturin D differs from iturin A by the difference of a free carboxyl group,[58] but the absolute structures of iturins D and E have not yet been fully established because of the limited amounts of metabolites available. The evidence suggests that in iturin D, one asparagine or glutamine that is present in iturin A is replaced by one asparagine or glutamate. Iturin E differs to iturin A by the presence of an asparagine-OCH_3 or glutamate-OCH_3. Both iturins D and E had potent activity against yeast and fungi with the following minimum

Figure 4 Structure of iturin A.

inhibitory concentrations (μg/ml): iturin D, *Saccharomyces cerevisiae* (15), *Candida albicans* (45), *Botrytis cinerea* (30), *Stemphylium radicinum* (45), and *Fusarium oxysporum* (30), while iturin E exhibited the following responses: *S. cerevisiae* (30), *C. albicans* (30), *B. cinerea* (10), *S. radicinum* (40), and *F. oxysporum* (40). These results readily point to the fact that modifications in the iturin A structure lead to changes in the specific activity of the molecule relative to the target. For example, in a field or storage situation clearly iturin E would be the chemical candidate likely to control *B. cinerea*. These minor changes in the iturin A structure, especially with respect to iturin E wherein the structure may consist of asparagine-OCH_3 or glutamate-OCH_3 (the OCH_3 perhaps being an important, though chemically minor modification), leads one to speculate that iturin A has a number of functions that may be altered synthetically and that these changes may well influence the biological activity relative to specific fungicidal activity. All these possibilities lend credence to the production of a broader spectrum fungicide using calculated iturin mixtures.

One of the most successful applications of iturin A in agriculture has been the use of a strain of *B. subtilis* B-3 in controlling brown rot, *Monilinia fructicola* (Wint.) Honey, as a postharvest treatment to peaches, nectarines, apricots, and plums. These experiments were eloquent in that they were one of the few in which a biological control agent and its biologically active constituent, a fungistatic product, were known. The *B. subtilis* was essentially the delivery system for the antibiotic, and the active ingredient in B-3 had unequivocally been shown to be iturin A.[59] The experiments had been instituted for two reasons. First, the benzimidazole fungicides were no longer efficacious because strains of *M. fructicola* had become resistant to them; and second, there is mounting public concern regarding the presence of pesticide residues in food. In addition, the future of the benzimidazole fungicides may be in jeopardy because of reregistration costs and other concerns. Furthermore, *B. subtilis* B-3 is readily obtained from liquid fermentation and iturin A has low toxicity.[60] In pilot tests on peaches B-3, when applied at rates of 2×10^8 to 7×10^8 colony-forming units per kilogram fruit was as effective as benomyl at 1 to 2 mg/kg fruit when incorporated into commercial coating wax. *In vivo* fungistatic activity was also noted when B-3 was applied to *M. fructicola, B. cinerea,* and *Glomerella cingulata.* This may have particular application in the grape industry to inhibit the noble rot of grape, *B. cinerea*; and in the kiwi growing areas of the world where *B. cinerea* is a problem.

B. NATURAL PRODUCT FUNGICIDES: BASIC DISCOVERIES AWAITING APPLICATION

A number of natural products have been isolated from fungi and plants that possess antifungal activity. While they have not yet been developed commercially, the potential exists for their use as specific target fungicides. Some of the fungally derived secondary metabolites are now discussed, especially with respect to their target specificity.

C. SPECIFIC ACTIVITY AND TARGET SPECIFICITY OF MICROBIAL METABOLITES

Natural product fungicides of the future may not have the broad spectrum exhibited by the synthetic fungicides, a luxury to which we have become accustomed, and treatment of a singular phytopathogen may have to be conducted on a case-by-case basis. However, by way of a trade-off, these new fungicides will be biodegradable. True, we have discussed the possibility of iturin mixtures for broadening the spectrum of activity, but for the most part natural product fungistatic agents tend to be quite specific in their activity. Nevertheless, when matched with the right target organism, their specific activity is high. There is, perhaps, no finer example of this than that of the antibiotic propanosine K-76 (Figure 5) against apple canker, *Valsa ceratosperma,* which is a major problem in

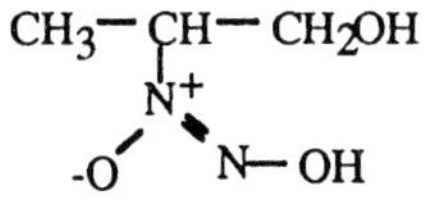

$$CH_3- CH- CH_2OH$$
$$N^+$$
$$-O \quad N- OH$$

Figure 5   Structure of propanosine K-76.

apple production in Japan. The pathogen gains its initial foothold in trees that have been wounded mechanically and in necrotic areas in the bark. There follows invasion of the trunk, and eventually the trees die. Propanosine K-76 was the result of screening microorganisms for possible antibiotic activity against *V. ceratosperma* and was isolated from an accessed culture, *Micromonospora chalcea* 671-AV2, as the water soluble sodium salt.[61] Consequently, this derivative is soluble in polar solvents such as water, methanol, and dimethyl sulfoxide, making for ease of application. While the salt is insoluble in organic solvents, this is easily rectified by making the nonsalt congeners for use in waxes and oil-based treatments. In terms of solvent range and application, the molecule is ideal. However, the most remarkable property of propanosine K-76 is its specificity (Table 2).

   The structure of the antibiotic was further proved by synthesis starting with 2-hydroxyaminopropanol. This was produced by reducing hydroxyacetone oxime, which consisted of both the E- and Z-isomers, with $NaBH_3CN$ in methanol for 3 h at pH 4. Only the E-isomer was easily reduced to 2-hydroxyaminopropanol; the Z-isomer was not. Following cleanup by chromatography, the compound was dissolved in 2NHCl; and $NaNO_2$ was added to finally give a racemic mixture of propanosine K-76. This mixture was only half as active as the natural product. That is, only one isomer was biologically active against *V. ceratosperma.* However, the molecule lends itself to synthetic exploitation, and presumably, the target specificity may be altered.

Table 2   **Activity of propanosine K-76 against bacteria and fungi**

Microorganism	Minimum inhibitory conc (μg/ml)
Colletotricum lagenarium	> 800
Fusarium oxysporum f. lycopersici	> 800
Gibberella fujikuroi	> 800
Pellicularia filamentosa	> 800
Saprolegnia parasitica	> 800
Alternaria kikuchiana	200
Botrytis cinerea	200
Cochliobolus miyabeanus	200
Diaporthe citri	200
Glomerella cingulata	200
Pyricularia oryzae	200
Rhizoctonia solani	200
Bacillus subtilis ATCC 6633	> 100
B. stearothermophilus	> 100
Candida albicans M 9001	> 100
Mycobacterium phlei 607	> 100
Staphylococcus aureus 209 P	> 100
Escherichia coli NIHJ	> 100
Pseudomonas aeruginosa M 8152	> 100
P. cepacia M 8242	> 100
Serratia marcescens	> 100
Vibrio percolens ATCC 8461	> 100
Saccharomyces cerevisiae Y21-1	25
Valsa ceratosperma	1

Adapted from Abe, Y. et al., *Agric. Biol. Chem.*, 47, 2703, 1983.

D. POTENTIAL NATURAL PRODUCT FUNGICIDES: THE FUTURE

There are several natural products possessing fungistatic or fungicidal properties that have yet to be commercially exploited, and some examples follow. These do not include antimicrobials that are used in the pharmaceutical industry as health care products and which, because of the relatively small amounts used in individual treatments, are very lucrative but have not yet been introduced for agronomic purposes. Nevertheless, the use of pharmaceutical fungicides in agriculture does pose a dilemma. Suppose, by way of a hyperbolic example, a medicinal fungicide costs $50 a pound to produce by fermentation; and after years of research and clearance by the various regulatory agencies, it is released on the market. Further suppose that the regimen for a patient is three, 100 mg tablets per day for 10 days at a cost of $2 per tablet; then the cost to a patient is $60. A total of 3 g of antibiotic have been ingested. The pound of fungicide has a potential gross income of $9080 by the time it reaches the patient. Assuming that the fungicide has agricultural potential and sells to the farmer for $100 per pound (again, allowing for 1 lb active ingredient per acre, per season), a problem ensues. The public, being somewhat savvy at times, could levy some difficult arguments about the cost of pharmaceuticals. However, there remains the possibility that those pharmaceutical fungicides to which certain human pathogens have become resistant may find a new use in the agricultural marketplace.

Among those natural products that warrant further screening for fungicidal activity are the koninginins.[62,65] These secondary metabolites were isolated from *Trichoderma koningii;* and while the compounds were discovered independently and simultaneously by two different research groups, the first of these to be tested against an economically important phytopathogen was 4,8-dihydroxy-2-(1-hydroxyheptyl)-3,4,5,6,7,8-hexahydro-2H-1-benzopyran-5-one (Figure 6) which proved to be effective *in vitro* against the take-all fungus of cereals, *Gaeumannomyces graminis* (Sacc.) Arx and Oliver var. *tritici* Walker.[62]

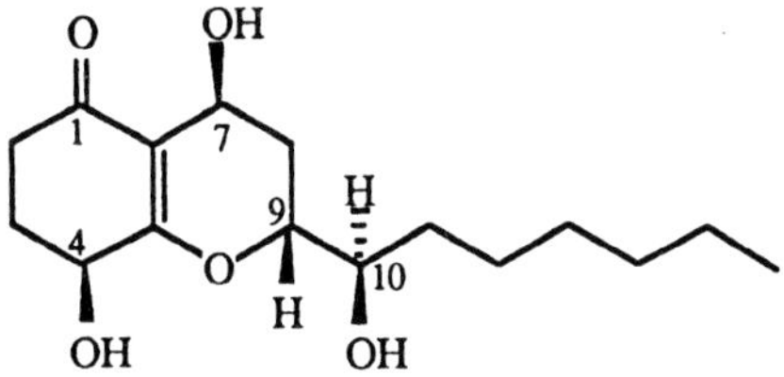

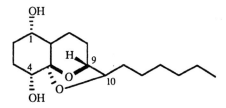

Figure 6 Structure of 4,8-dihyxdroxy-2-(1-hydroxyheptyl)-3,4,5,6,7,8-hexahydro-2H-1-benzopyran-5-one.

The organism had been isolated from soil that was found to be suppressive to the growth of the take-all fungus, and the sleuthing involved from the initial observation to the final chemical solution of the responsible antibiotic was formidable. At the same time, koninginin A (Figure 7) was isolated from *T. koningii* found in soil in which a wilting *Diffenbachia* plant was growing.[67] Surprisingly, koninginin A was inactive against *Curvularia lunata*-49, *Aspergillus flavus* Link ex Fr., and *Chaetomium cochlioides*-195 at concentrations up to 500 μg/4 mm disk in petri dish assays. However, koninginin A was later isolated from *T. harzianum*, strain 71, and demonstrated activity against *G. graminis* at 250 μg.[64] Koninginin B has yet to be tested against the take-all fungus.[65]

Figure 7 Structure of Koninginin A.

In addition to the koninginins isolated from *T. harzianum,* another accession (MI 311092) — strain 73 — has yielded two new butenolide structures (each possessing a 3,4-dialkylfuran-3(5H)-one nucleus) that were active against *G. graminis*. These are 3-(2-hydroxypropyl)-4-(hexa-2E,4E-dien-6-yl)furan-2-(5H)-one (Figure 8)[64] and 3-(propenyl)-4-(hexa-2E,4E-dien-6-yl)furan-2(5H)-one (Figure 9).

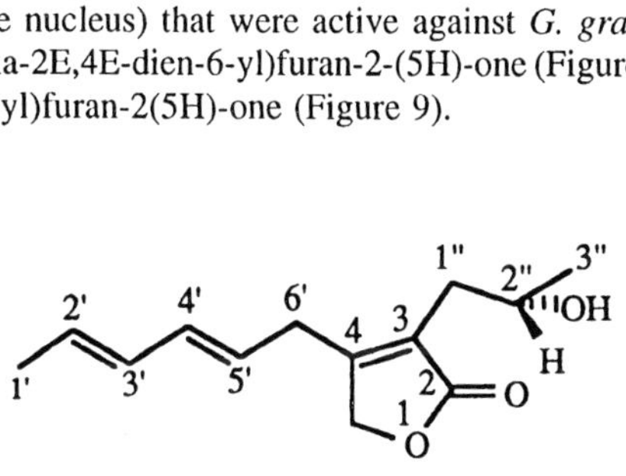

Figure 8 Structure of 3-(2-hydroxypropyl-4-(hexa-2E,4E-dien-6-yl)furan-2(5H)-one.

148

Figure 9 Structure of 3-(propenyl)-4-(hexa-2E,4E-dien-6-yl)furan-2(5H)-one.

Trichoderma harzianum, as we have seen, is already in experimental use as a biocontrol agent. Likewise, *T. koningii* now becomes a viable candidate for screening as another biocontrol agent. Also, each of the fungistatic or fungicidal natural products isolated from the Trichodermas is ready for chemical elaboration by synthesis or biotransformation.

Another set of natural products has also been used to control the take-all fungus of cereals *in vitro.* They are terrein (Figure 10) and E-4-(1-propen-1-yl)-cyclopenta-1,2-diol (Figure 11), both metabolites of *Aspergillus terreus.* It should be noted that *A. terreus* was toxic to most wheat and rye grass seedlings when inoculated at 1%, but at 0.5% this was not the case. At the 0.5% rate of application, *A. terreus* was antagonistic to the take-all fungus when incorporated in sterilized and unsterilized soil.[66] This is an interesting example of where, if it is successful, a dose of a biocontrol agent must be critically measured prior to field application, and concomitantly the titer of a biologically active ingredient must be absolutely known in any given batch.

Figure 10 Structure of terrein.

Figure 11 Structure of E-4-(1-propen-1-yl)-cyclopenta-1,2-diol.

Nature is remarkably erudite and biosynthesizes molecules, which would try the patience and skills of most chemists, with great facility. Repetitively one sees hauntingly familiar themes being played in novel ways. Such is the case with the gamahonolides (Figure 12) obtained as novel antifungal compounds from the stroma of *Epichloe typhina* on *Phleum pratense,* which are chemical relatives of 6-pentyl-α-pyrone. Simply, *Epichloe typhina* is a phytopathogenic fungus that causes choke disease in timothy grass, *Phleum pratense.* When the organism infects timothy, the grass acquires resistance to secondary pathogens, especially leaf spot, *Cladosporium phlei.* This observation led to the eventual isolation of gamahonolide A and B, gamahorin, and 5-hydroxy-4-phenyl-2(5H)-furanone.[67] The structures for these are shown in Figures 12, 13, and 14.

The antifungal properties of the metabolites were evaluated against *Cladosporium herbarum* using a thin-layer chromatography plant technique. Gramahonolide A inhibited

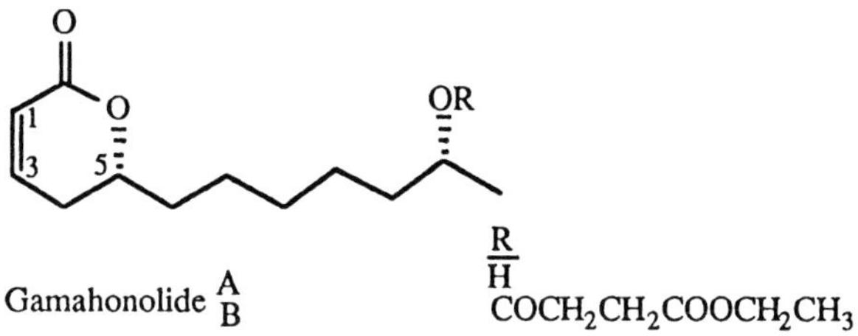

Figure 12   Structure of gamahonolides A and B.

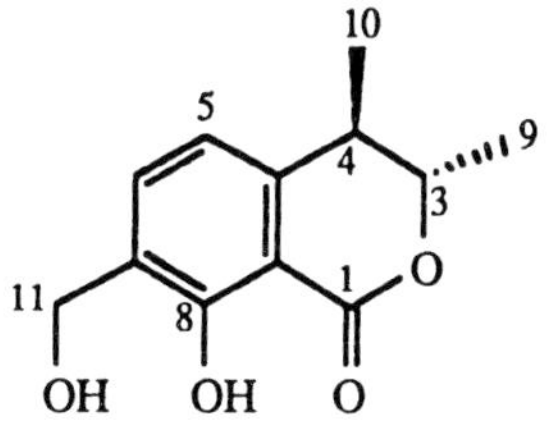

Figure 13   Structure of gamahorin.

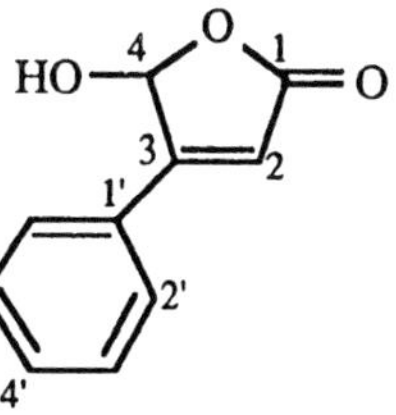

Figure 14   Structure of 5-hydroxy-4-phenyl-2(5H)-furanone.

C. herbarum with 25 µg per spot while gamahorin and 5-hydroxy-4-phenyl-2(5H)-furanone only required 10 µg per spot. Gamahonolide B was not assayed but appeared to have the same inhibitory properties as gamahonolide A based on bioassays run during fractionation of the compounds.

These secondary metabolites represent only a small portion of those isolated that have antimicrobial activity.

VI.   CONCLUSIONS

The answers as to how natural fungicides will replace synthetic ones are like a puzzle that has yet to be pieced together. Bits of the puzzle already exist. For example, we know that certain biocontrol microorganisms efficiently control phytopathogens and that a number of natural products also control pernicious bacteria and fungi. In some cases, both the biocontrol agents and their biologically active products are known so that a delivery system capable of producing a fungistatic or fungicidal agent in a continuum, as opposed to the application of a chemical bolus, can be used. The natural products themselves are open to chemical elaboration to produce other fungicides, thereby increasing the useful range of the product. Perhaps what is needed more than anything else is a change in philosophy. The fungicides of the future will be highly target specific, have high specific activity, and will be biodegradable. All these factors require a radical change in thinking and approach if we are to be successful. Rest assured that there are myriad biocontrol agents and fungicidal metabolites yet to be discovered in nature's storehouse.

REFERENCES

1. **Bronowski, J.,** *The Ascent of Man*, Little, Brown & Company, Boston, 1971.
2. **Kurosawa, E.,** Experimental studies on the secretion of *Fusarium heterosporum* on rice plants, *J. Nat. Hist. Soc. Formosa*, 16, 213, 1926 (Engl. translation in *Source Book of Gibberellin* 1828–1957, Stodola, F. H., Ed., Agricultural Research Service, U.S. Department of Agriculture, ARS-71-11, 1958).
3. **Cutler, H. G. and Schneider, B. A.,** in *Plant Growth Regulator Handbook of the Plant Growth Regulator Society of America*, 3rd ed., 1990, 69.
4. **Hayashi, T.,** in *Plant Growth Regulation*, The Iowa State University Press, Ames, IA, 1961.
5. **Szekely, E. B.,** in *The Ecological Health Garden. The Book of Survival*, International Biogenic Society, San Diego, CA, 1978.
6. **Curl, E. A.,** *CRC Rev. Plant Sci.*, 7, 175, 1988.
7. **Hill, R. A.,** in *Proc. Practical Development and Implementation of Biological Control as Agents for Pest and Disease Control* Workshop and Lectures, Canterbury Agricultural Centre (Lincoln), New Zealand, 1989.
8. **Lumsden, R. D., Garcia-E. R., Lewis., and Frias-T., G. A.,** in *Agroecology: Researching the Ecological Basis for Sustainable Agriculture*, Springer-Verlag, New York, 1990, 464.
9. **Papavizas, G. C. and Lewis, J. A.,** in *Biological Control in Crop Production*, BARC Symp. No. 5, Beltsville, MD Allanheld, Osmun Publishers, Totowa, NJ, 1981.
10. **Broadbent, P. and Baker, K. F.,** in *Biology and Control of Soil-borne Plant Pathogens*, American Phytopathological Society, St. Paul, MN, 1975.
11. **Hill, R. A.,** *Horticultural Produce and Practice*, HPP93, Ministry of Agriculture and Fisheries, Wellington, New Zealand, 1988.
12. **Leach, R.,** *Proc. R. Soc. of London, Soc. Ser. B.*, 121, 56, 1936.
13. **Van der Pas, J. B., Hood, I. A., and MacKenzie, M.,** *Forest Pathology in New Zealand* No. 4, 1983.
14. **Birch, T. T. C.,** *N. Z. For. Serv. Bull.*, 9, 1937.
15. **Swift, M. J.,** *Forestry*, 45, 67, 1972.
16. **Shaw, C. G., III,** Epidemiological Insights into *Armillaria mellea* Root Rot in a Managed Ponderosa Pine Forest, Ph.D. thesis, Oregon State University, Corvallis, OR, 1975.
17. **Shaw, C. G., III and Calderon, S.,** *N. Z. J. For. Sci.*, 7, 359, 1977.
18. **Kile, G. A.,** The significance of *Armillaria* species in eucalypt dieback, *Proc. Conf. CSIRO Div. For. Res.*, Canberra, Australia, 1980.
19. **Benjamin, M.,** Studies on the Biology of *Armillaria* in New Zealand, Ph.D. thesis, University of Auckland, New Zealand, 1983.
20. **Hood, I. A. and Sanberg, C. J.,** *Proc. 7th Int. Conf. Root Butt Rots*, Vernon and Victoria, British Columbia, Canada, 1989.
21. **Hill, R. A.,** A Report to Tasman Forestry Ltd., Rotorua, New Zealand, 1991.
22. **Hill, R. A.,** *Proc. New Zealand Kiwifruit Marketing Board, National Research Conf.*, Rotorua, New Zealand 20, 1990.
23. **Corke, A. T. K. and Hunter, T.,** *J. Hortic. Sci.*, 54, 47, 1979.

24. Greer, J. E., Antagonistic Reactions of *Trichoderma harzianum* toward *Rhizoctonia solani* and *Sclerotium rolfsii*, M.Sc. thesis, University Georgia, Athens, GA, 1978.
25. Truong, H. X., Salinas, M. D., Obien, A. S., and Carasi, R. C., *Proc. Brighton Crop Prot. Conf. — Pests and Diseases*, Brighton, U.K., 1988.
26. Elad, Y., Chet, I., and Katan, J., *Phytopathology*, 70, 119, 1980.
27. Chet, I., Elad, Y., Kalfon, A., Hadar, Y., and Katan, J., *Phytoparasitia*, 10, 229, 1982.
28. Strashnov, Y., Elad, Y., Sivan, A., Rudich, Y., and Chet, I., *Crop Prot.*, 4, 359, 1985.
29. Lewis, J. A. and Papavizas, G. C., *Phytopathology*, 70, 85, 1980.
30. Chet, I., in *Innovative Approaches to Plant Disease Control*, Wiley Interscience, New York, 1989.
31. Abd-El-Moity, T. H. and Shatla, M. N., *Phytopathol. Z.* 100, 29. 1981.
32. Elad, Y., Zviell, Y., and Chet, I., *Crop Prot.*, 5, 288, 1986.
33. Marois, J. J., Mitchell, D. J., and Sonoda, R. M., *Phytopathol. A,* 71, 1257, 1981.
34. Sivan, A., *Plant Dis.*, 71, 587, 1987.
35. Dutta, B. K., *Plant Soil*, 63, 209, 1981.
36. Dye, M. H., Silverleaf Disease of Fruit Trees, New Zealand Ministry of Agriculture and Fisheries Bulletin No. 104, 1972.
37. Grosclaude, C., Ricard, J., and Dubos, B., *Plant Dis. Rep.*, 57, 25, 1973.
38. Dubos, B. and Ricard, J. L., *Plant Dis. Rep.*, 58, 147, 1974.
39. Tronsmo, A. and Raa, J.,*Phytopathol. Z.*, 89, 215, 1977.
40. Papavizas, G. C., *Annu. Rev. Phytopathol.*, 23, 23, 1985.
41. Taylor, H., Laboratory Studies on the Interactions between *Armillaria novae zelandiae* and *Trichoderma* Species, M.Sc. thesis, University of Auckland, New Zealand, 1991.
42. Cutler, H. G., Crumley, F. G., and Cole, P. D., *Agric. Biol. Chem.*, 50, 2943, 1986.
43. Claydon, N., Allen, M., Hanson, J. R., and Avent, A. G., *Trans. Br. Mycol. Soc.*, 88, 503, 1987.
44. Godtfredsen, W. D. and Vangedal, S., *Acta Chem. Scand.*, 19, 1088, 1965.
45. Cutler, H. G. and LeFiles, J. H., *Plant Cell Physiol.*, 19, 177, 1978.
46. Cutler, H. G., in *The Science of Allelopathy*, John Wiley & Sons, New York, 1986.
47. Majumdar, S. K. and Bose, S. K., *Nature (London)*, 181, 134, 1958.
48. Sengupta, S. and Bose, S. K., *Biochem. Biophys. Acta*, 237, 102, 1971.
49. Ebata, M., Miyazake, K., and Takahashi, Y., *J. Antibiot.*, 22, 467, 1969.
50. Peypoux, F., Michel, G., and Delcambe, L., *Eur. J. Biochem.*, 63, 391, 1976.
51. Besson, F., Peypoux, F., Michel, G., and Delcambe, L., *Eur. J. Biochem.* 77, 61, 1977.
52. Korzybski, T., Kowszyk-Gindifer, Z., and Kurylowicz, W., in *Antibiotics: Origin, Nature and Properties*, Vol. 3, American Society for Microbiology, Washington, D.C., 1978, 1529.
53. Isogai, A., Takayama, S., Murakoshi, S., and Suzuki, A., *Tetrahedron Lett.*, 23, 3065, 1982.
54. Besson, F. and Michel, G., *J. Antibiot.*, 40, 437, 1987.
55. Peypoux, F., Guinand, M., Michel, G., Delcambe, L., Das, B. C., Varenne, P., and Lederer, E., *Tetrahedron*, 29, 3455, 1973.
56. Loeffler, W., Tschen, J. S.-M., Vanittanakom, N., Kugler, M., Knorpp, E., Hsieh, T.-F., and Wu, T.-G., *J. Phytopathol.*, 115, 204, 1986.
57. Phae, C. G., Shoda, M., and Kubota, H., *J. Ferment. Bioeng.*, 69, 1, 1990.
58. Peypoux, F., Besson, F., Michel, G., Delcambe, L., and Das, B. C., *Tetrahedron*, 34, 1147, 1978.
59. Gueldner, R. C., Reilly, C. C., Pusey, P. L., Costello, C. E., Arrendale, R. F., Cox, R. H., Himmelsbach, D. S., Crumley, F. G., and Cutler, H. G., *J. Agric. Food Chem.*, 36, 366, 1988.

60. Pusey, P. L., *Pest. Sci.,* 27, 133, 1989.
61. Abe, Y., Kadokura, J.-I., Shimazu, A., Seto, H., and Otake, N., *Agric. Biol. Chem.,* 47, 2703, 1983.
62. Dunlop, R. W., Simon, A., and Sivasithamparam, K., *J. Nat. Prod.,* 52, 67, 1989.
63. Cutler, H. G., Himmelsbach, D. S., Arrendale, R. F., Cole, P. D., and Cox, R. H., *Agric. Biol. Chem.,* 53, 2605, 1989.
64. Almassi, F., Ghisalberti, E. L., and Narbey, M. J., *J. Nat. Prod.,* 54, 396, 1991.
65. Cutler, H. G., Himmelsbach, D. S., Yagen, B., Arrendale, R. F., Jacyno, J. M., Cole, P. D., and Cox, R. H., *J. Agric. Food Chem.,* 39, 977, 1991.
66. Ghisalberti, E. L., Narbey, M. J., and Rowland, C. Y., *J. Nat. Prod.,* 53, 520, 1990.
67. Koshino, H., Yoshihara, T., Okuno, M., Sakamura, S., Tajimi, A., and Shimanuki, T., *Biosci. Biotechnol. Biochem.,* 56, 1096, 1992.
68. Yabuta, T. and Sumiki, Y., *J. Agr. Chem. Soc. Japan,* 14, 1526, 1938.

Chapter 13

Manipulation of Defense Systems with Elicitors to Control Postharvest Diseases

Ahmed El Ghaouth

CONTENTS

I. INTRODUCTION

Fungal diseases represent a major limiting factor in the long-term storage of fruit and vegetables.[1] In developing nations losses associated with fungal and bacterial diseases have been estimated roughly at 40 to 50%. While in developed nations losses may be lower, they are often just as serious from the producer's and consumer's standpoint. Postharvest infection can occur either prior to harvest or during harvesting and subsequent handling and storage. Disease development during the postharvest phase depends on the physiological status of the tissue and the constitutive and inducible resistance mechanisms of the harvested produce. In general, most harvested commodities are resistant to fungal infection during their early postharvest phase. However, during ripening and the senescence phase, they become more susceptible to infection.[2]

Currently, synthetic fungicides constitute a primary means of controlling postharvest diseases.[3] However, public and scientific concerns about the presence of synthetic chemicals in the food chain have generated an interest in the development of alternative approaches for the control of storage diseases. Recently, use of biological antagonists has been advanced as an alternative approach to chemical therapy; and several patented antagonistic microorganisms are near commercialization.[4,5] Although fruits and vegetables possess constitutive and inducible defense mechanisms which enable them to ward off infection, this potential has not been fully explored as a manageable form of resistance. This may have stemmed from our limited understanding of the nature of defense mechanisms in harvest commodities and the way they might be manipulated to enhance resistance. We often perceive harvested commodities as highly perishable products with reduced abilities to express defense responses to infection. Most of our knowledge concerning inducible resistance in harvested commodities is inferred from results obtained with vegetative tissue where induced resistance is recognized as an important

0-8493-4567-7/94/$0.00+$.50

form of crop protection.[6,7] The purpose of the present chapter is to introduce an emerging form of postharvest disease control which involves the intensification of natural defense mechanisms through the use of elicitors. The fundamental basis and the potential of this novel approach will be discussed with special reference to UV-C and chitosan treatments.

II. HOST DEFENSE MECHANISMS: OVERVIEW OF BIOCHEMICAL ASPECTS

It is well established that plants, as other living systems, are able to activate several cellular processes which protect them from invasion. These biochemical and structural processes are part of an integrated set of highly coordinated resistance responses that are triggered upon infection or treatment with elicitors and help ward off the spread of pathogens.[8-11] On the recognition of a potential pathogen or its products, plant tissue usually responds by producing several local and systemic responses. Localized protective responses are expressed in tissue around the site of interaction and may include rapid death of the cells in the immediate vicinity of the infection site, referred to as a hypersensitive reaction (HR), callose formation, lignification, and phytoalexin accumulation. By contrast, systemic defense responses develop in cells distant from the site of the stimulus and involve the activation of antifungal hydrolases such as chitinase, β-1,3-glucanases, and peroxidases. Both localized and systemic responses, when expressed soon enough and in sufficient magnitude, can have a deterrent effect on disease development. While the changes associated with resistance are dependent on the interaction of genetic factors in both the host and pathogen, they can be triggered by treatment with biotic and abiotic elicitors.[12,15] Several biologically active elicitors have been identified.[14,15] The use of elicitors, in particular fungal wall glucans, has enhanced our understanding of the transcriptional activation of defense genes.[9-11]

In recent years, considerable attention has been placed on induced resistance in vegetative crops as an important manageable form of plant protection.[6,7,16,17] This interest has been generated by the large volume of basic and applied research which demonstrates that plants can be rendered resistant by artificially turning on their natural defense mechanisms.[7,16] Various types of induced resistance have been reported in different host/pathogen systems.[7,16-18] The work of Kuć and collaborators[6,7] bears witness to the validity of induced resistance as a novel and promising approach for disease control. Attempts to exploit induced resistance, either by the development of transgenic plants that constitutively express inducible defense genes or through the application of defense response elicitors, are actively being made in many laboratories. While the phenomenon of inducible resistance can also be manipulated in harvested commodities, it has received little attention until recently. The control of postharvest diseases through treatment with fungal cell wall fragments,[19] chitosan,[20] and UV-C light[21,22] strongly suggests that an activation of defense responses in harvested crops is feasible and may offer a new strategy for disease control.

Inducible resistant responses in harvested tissue may be comparable in many facets to that in vegetative tissue. However, since in harvested commodities the losses of respirable substrates are not made up for their ability to initiate defense responses, an energy demanding process, is likely to decline with ripening. Indeed, the resistance of fruits and vegetables diminishes rapidly with the onset of ripening. This is well illustrated by the resumption of the activity of quiescent infections during the ripening phase. Therefore, any strategy aimed at exploiting the defense potential of fruits and vegetables ought to take into consideration the physiological status of the tissue and the external factors that affect its physiology.

Among the diverse biochemical defense responses, deliberate stimulation of hydrolytic enzymes, induction of structural barriers, and accumulation of phytoalexins to an inhibitory level could be potential methods for enhancing resistance in harvested tissue. In this chapter the discussion will be limited primarily to cellular processes involved in the stimulation of antifungal glucanohydrolases and the accumulation of phytoalexins.

A. ANTIFUNGAL GLUCANOHYDROLASES

Glucanohydrolases such as chitinase, chitosanase, and β-1,3-glucanase are low molecular proteins that hydrolyze the main components of fungal cell walls.[10,13,23] These enzymes have received considerable attention because they are considered to play a major role in constitutive and inducible resistance of plants against invading pathogens. There is good evidence that the action of endoglucanohydrolases results in the inhibition of fungal growth,[24,25] and in the release of signaling molecules (β-glucan, chitin, and chitosan oligomers) that further induce defense reactions.[26,27] Chitinases and β-1,3-glucanases have been found in monocots and dicots, and are known to be stimulated by infection in response to treatment with elicitors. For a full discussion, the reader is referred to reviews.[10,13,23,28]

Glucanohydrolase activity and genetic regulation have been studied primarily in stressed vegetative tissue and in host-pathogen interactions.[10,23] In plant tissue, it has been suggested that antifungal hydrolases provide a dual mechanism of defense. The extracellular hydrolases constitute the first line of defense; intracellular ones constitute the second, when plant cell membranes are breached.[29] The data obtained in induced resistance studies add another dimension not related to their functionality, but instead to their long-term and generalized protection.[7,16]

Even though these enzymes with antifungal potential are probably present and may contribute to the resistance of postharvest tissue, we have often failed to acknowledge their potential in disease control. Activating and maintaining glucanohydrolases in harvested tissue through prestorage treatment with elicitor(s) could be a promising means of enhancing disease resistance and consequently prolonging storage life. The systemic nature and persistence of these enzymes in plant tissue upon elicitation could be of significant importance in retarding the resumption of the activity of quiescent infections which become active when tissue resistance declines.

B. PHYTOALEXINS

Phytoalexins are antimicrobial secondary metabolites which are synthesized *de novo* and accumulate in plants in response to infection and treatment with elicitors.[30,31] They show a wide diversity in structure and biogenetic origin and are predominately phenylpropanoids, isoprenoids, and acetylenes.[31] There are numerous reviews on phytoalexins and their mode of action. Investigation into the implication of phytoalexins in disease resistance has dominated most studies of active defense responses. A large body of evidence supports a disease resistance function for phytoalexins (see Bailey and Mansfield[31]). Bailey and Mansfied,[31] Stoessel,[32] and Haard and Cody[33] have summarized the known phytoalexins in various crop species. At least 100 plant species representing 21 families have been shown to accumulate phytoalexins.

Despite much progress in isolating and characterizing phytoalexins in major crop plant, information regarding postharvest commodities is still limited. Very little is known about the effects of harvest, postharvest environments and maturity on synthesis and accumulation of phytoalexins in fruits and vegetables. Recently, Creasy and Coffee[34] showed that phytoalexin response decreased with maturity in grape berries. In several postharvest commodities the resumption in activity of latent infection is believed to be associated with a decreased ability of tissue to synthesize phytoalexins and preformed inhibitors.[2,35] In light of this and the fact that phytoalexins are antimicrobial, it is reasonable to assume that deliberate induction of phytoalexins may give postharvest tissue an advantage in fighting infection.

III. ELICITORS FOR THE CONTROL OF POSTHARVEST DISEASES

In recent years, there have been several attempts to manipulate natural defense mechanisms of harvested commodities by elicitors to control postharvest decay. This approach may hold promise as a viable alternative to chemical therapy. Its time is also ripe in that all over the world the use of synthetic fungicides in agriculture is under scrutiny. The recent work of Adikaram et al.,[19] El Ghaouth et al.,[20] Mercier et al.,[21] and Stevens et al.[22] provide evidence that activation of defense responses by prestorage treatment with elicitors is feasible and has potential in reducing postharvest diseases.

IV. CONTROL OF POSTHARVEST DISEASES WITH ULTRAVIOLET LIGHT

Nonionizing UV-C (from 190 to 280 nm) radiation is known to affect several biological processes in plant tissue[36,37] including the induction of anthocyanin, flavonoid biosynthesis,[38,39] and phytoalexins.[34,40,41] Elicitation of phytoalexins by UV-C has been reported in several crops and was shown to increase resistance of plant seedlings and hypocotyls.[41,42] Recently, Stevens and collaborators[22] showed that treatment of onions, sweet potatoes, tomatoes, peaches, and citrus fruits with hormetic low doses of UV-C reduced disease incidence and delayed ripening. In mature green tomato fruit, UV-C treatment slowed the rate of color change and increased flesh firmness. Similar delays in ripening have also been reported by Arul et al.[43] in UV treated bell peppers and tomatoes and is thought to stem from an increase in antioxidants and free-radical scavengers.

Reduction of postharvest decay by UV-C treatment has also been reported by other researchers and was attributed to the ability of the treatment to induce antifungal secondary metabolites instead of to its germicidal effect.[21,44,45] For instance, in UV-treated citrus fruit the onset of induced resistance was found to coincide with the induction of phenylalanine ammonia-lyase (PAL) activity, a key enzyme in the phenylpropanoid pathway.[44] In carrot slices, the induction of 6-methoxymellen by UV-C was shown to increase the resistance of the tissue to *Botrytis cinerea* and *Sclerotinia sclerotiorum* infection.[21] Similarly, a relationship between the level of scoparone and resistance to *Penicillium digitatum* has been reported in UV-treated lemon fruit.[45]

Although a role of phytoalexins in induced resistance by UV treatment has often been inferred, it has not been substantiated by experimental evidence. Detailed studies are needed to determine whether other defense responses apart from phytoalexin accumulation are involved in the observed protection by UV treatment. Such information would allow the development of a more effective strategy aimed at exploiting the potential of UV treatments.

While the potential of UV treatment as a disease control method has been demonstrated with several commodities, its applicability may be hampered by the fact that: (1) its protective effect often does not persist for long periods of time and (2) the responsiveness of the harvested tissue is likely to decline during the ripening process. Thus, UV treatment by itself is unlikely to provide disease control comparable to that obtained with synthetic fungicides. For UV treatments to gain a broader acceptance, their protective effect needs to be maintained for a significant storage period. This could be achieved by combining UV treatments with other promising alternatives such as antagonistic microorganisms and naturally occurring biocides. Indeed, preliminary studies have shown that the combination of UV treatments with antagonistic yeasts enhances and prolongs the protection of fruit against postharvest pathogens.[45a]

Prior to any attempt to use UV treatment, it is important to determine whether it results in the accumulation of unpalatable secondary metabolites by activating aromatic or terpenoid biosynthesis. If the accumulation of undesirable compounds occurs, this will outweigh its protective potential and further deter its application.

V. CHITOSAN: A POTENTIAL PRESERVATIVE FOR POSTHARVEST COMMODITIES

Another alternative approach that has been actively pursued involves the use of a natural bioactive compound, chitosan, that interferes with fungal growth and activates defense mechanisms in the plant tissue. This dual functionality gives chitosan potential as an antifungal preservative for fresh horticultural commodities. Chitosan is a β-1,4-glucosamine polymer that is found as a natural constituent in the cell wall of many fungi.[46] It is produced from chitin of arthropod exoskeletons that has been deacetylated to provide sufficient free amino groups to render the polymer readily soluble in diluted organic acids.[47] This polymer is known to interfere with growth of a wide range of fungi,[20,27,48] to induce a multitude of biological processes in host tissue,[20,27,49,50] and to form a semipermeable film.[51-53] Furthermore, chitosan, a by-product from the seafood industry, appears to be safe as indicated by toxicological studies.[54]

In recent years, there has been an increasing interest in the use of chitosan as a specialty chemical in several fields including agriculture, and several potential biological functions have been identified.[55-57] Attempts to exploit the filmogenic and biological properties of chitosan were recently made with several postharvest commodities. Chitosan, when applied as a coating was found to delay ripening by acting as a barrier to gas diffusion,[58-60] reduce the incidence of decay, and stimulate several defense responses in plant tissue.[20,61] Coating tomato, cucumber, bell pepper, and strawberry fruits with chitosan reduced their respiration rate and weight loss, improved their appearance, and extended their shelf life.[58] Coated fruits were firmer and higher in titratable acidity than control fruits, indicating an overall delay of ripening. In tomato fruit, delay of ripening by chitosan coating was associated with its ability to modify the internal atmosphere.[59] Coating provided a greater barrier for the efflux of CO_2 than for the influx of O_2. While instances of interference with normal fruit ripening have been reported with other edible coatings,[52,62] chitosan coating neither altered the ripening capacity of the fruit nor caused any apparent phytotoxicity.[58] In addition, chitosan, unlike most available coating materials that mimic the beneficial effect of modified atmosphere, offers the added advantage of reducing postharvest decay.

A. ANTIFUNGAL ACTIVITY

When applied as a coating, chitosan was effective in reducing decay of tomatoes, bell peppers, cucumbers, and strawberries caused by *B. cinerea* and *R. stolonifer*. In strawberry fruit stored at 13°C, a chitosan treatment was as effective as the fungicide Rovral in controlling decay caused by *B. cinerea*. There was no difference between chitosan and Rovral treatments in controlling decay up to 21 d of storage. Thereafter, Rovral-treated berries decayed at a higher rate than chitosan-treated ones. Reduction of decay by chitosan was also observed in tomato and bell pepper fruits stored at 20°C.[63] Chitosan, when applied as a stem scar treatment of bell peppers, markedly reduced lesion development caused by *B. cinerea*.[63,64] By the end of a 14-d storage period, less than 20% of the chitosan-treated fruits developed small lesions, whereas all the inoculated control fruits had reached an advanced stage of decay. The observed reduction of lesion development by chitosan appears to originate in part from its antifungal property. *In vitro* studies revealed that chitosan is effective in reducing the growth of *B. cinerea*, *Colletotrichum gloeosporioides*, and *R. stolonifer*.[20,65] The inhibitory activity of chitosan increases with concentration and the degree of deacetylation.[65] While the mechanism underlying the antifungal action of chitosan is not well understood, it seems to consist of more than one mode. Chitosan was found to cause cellular leakage and morphological alterations in *Botrytis cinerea* and *Rhizopus stolonifer*.[65] Chitosan treatment induced severe morphological alterations characterized by excessive branching and wall alteration in *R. stolonifer* presumably via its action on a cell wall-forming enzyme, namely, chitin deacetylase.[65,66]

158

Ultrastructural and cellular alteration exhibited by *B. cinerea*[64] and *Fusarium oxysporum*[57] grown in the presence of chitosan further suggests that chitosan exerts its inhibitory activity by more than one single mode. The nature of the mechanisms by which chitosan affects fungal cell wall biosynthesis and the implication of such alteration on the ability of the pathogen to initiate infection merit further investigation.

B. CYTOCHEMICAL ASPECTS

Although there is considerable literature on the biological activity of chitosan, the basis of the mechanism that underlies its ability to increase host resistance and thereby prevent disease development has not been fully elucidated. In order to maximize its potential use as an antifungal preservative an understanding of how chitosan exerts its effect on disease development *in planta* is necessary. Studies conducted with bell peppers revealed that

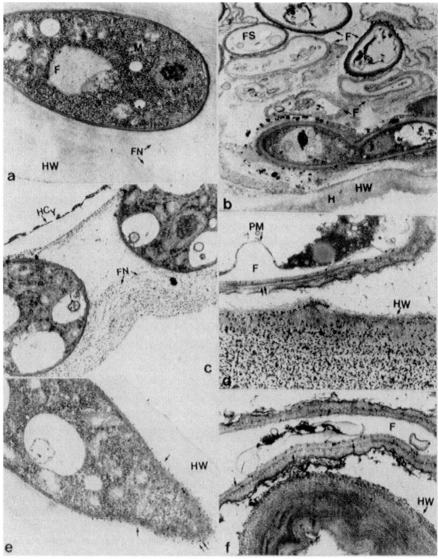

Figure 1

chitosan, when applied as a stem scar treatment, controlled disease development and caused severe damage to *Botrytis cinerea*.[64] Examination of several sections from bell pepper fruit inoculated with *B. cinerea* showed that fungal colonization proceeded rapidly causing extensive cellular disorganization. Infection hyphae ramified intercellularly and intracellularly as well as within the host wall (Figure 1a), causing marked degradation of pectin and disruption of the structural layering of cellulose of the host walls (Figure 1c). During their ingress, hyphal cells also secreted macerating enzymes extracellularly as indicated by degradation of pectic molecules in wall areas distant from the site of infection.[64] The capability of *B. cinerea* to secrete macerating enzymes, in particular pectinolytic enzymes, is believed to be of primary importance in disease development, since they can enhance the permeation of host metabolites and suppress active host resistance responses. Invading hyphae appeared metabolically active and showed no apparent signs of damage during their ingress in the host tissue (Figures 1a and c).

In chitosan-treated tissue, however, invading hyphae were predominantly found in wound cavities and surrounding ruptured cells.[64] Although few of the invading hyphae appeared normal, most displayed severe cellular disorganization that ranged from vacuolation to complete protoplasm disintegration (Figure 1b and Figure 2a-d). Pathogen ingress in the cell layers beneath the wounded area was not accompanied by marked host cell disorganization. Host primary walls, even when closely appressed against fungal cells appeared well preserved and showed no obvious sign of alteration of pectin and cellulose distribution. The results obtained with the exoglucanase-gold complex showed that labeling of cellulose was intense and regular over host walls in contact with fungal cells (Figure 1d). The preservation of pectic and cellulosic binding sites in the host wall strongly suggests that chitosan might have affected the capability of *B. cinerea* to produce macerating enzymes. Indeed, examination of several sections revealed that fungal cells have severe ultrastructural and cellular damage as a result of chitosan treatment (Figure 2a-d). Such highly altered fungal cells are unlikely to be able to initiate successful infection.

In fungal cells highly altered by chitosan treatment, the plasma membrane tended to retract from the cell wall, and in many cases the cytoplasm assumed a very electron-dense appearance (Figure 2a and b). The retraction of the plasma membrane was often followed by the deposition of chitin-rich material in paramural spaces, suggesting that

Figure 1 Transmission electron micrographs of chitosan-treated and non-treated bell pepper fruit tissue inoculated with *Botrytis cinerea:* (**a, c,** and **e**) Inoculated control; (**b, d** and **f**) chitosan treated. (**a**) Pathogen growth within host wall was followed by pronounced alteration and swelling of the cell wall. Note the fibrillar network appearance of shredded wall (arrows). (**c** and **d**) Labeling with the gold complexed exoglucanase for localization of cellulosic β-1,4-glucan. (**c**) Inoculated control, pathogen ingress in host wall is followed with complete disruption of the wall fibrillar structure(arrows). Labeling was completely abolished over localized areas of the wall. (**d**) Chitosan-treated, intense and regular labeling over host walls facing fungal cells. (**e** and **f**) Labeling with the WGA/ovomucoid-gold complex for localization of chitin. (**e**) Under control, fungal growth within the host wall was not followed by alteration in chitin-labeling pattern. Labeling was evenly distributed over the wall of invading hyphal cell. (**f**) Chitosan-treated, labeling discontinuities are observed over fungal walls facing host walls. Several delineated fungal wall areas appear devoided of labeling. **Abbreviations:** (AM) amorphous material; (Cy) cytoplasm; (DM) deposited material; (F) fungal cell; (FM) fibrillar material; (FN) fibrillar network; (FS) fungal shell; (FW) fungal wall; (HW) host cell wall; (HCy) host cytoplasm; (IS) intercellular space; (P) papilla; (M) mitochondria; (OM) opaque material; (PM) plasma membrane; (S) septum; (Va) vacuole; (Ve) vesicle.

chitosan may have increased *de novo* synthesis of chitin (Figure 2c and d). The biological significance of the newly formed material may well be to counterbalance the turgor pressure created by the alteration of wall and the gradual retraction of the plasma

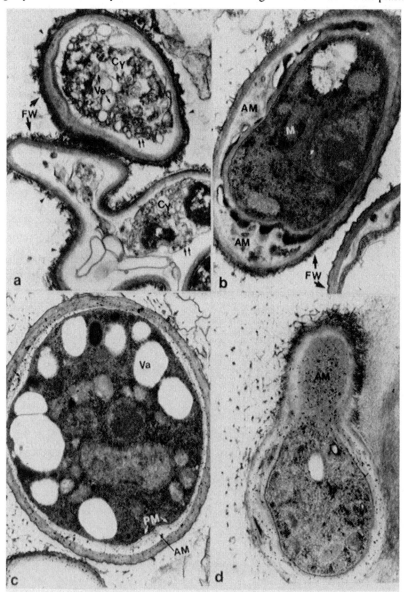

Figure 2 Transmission electron micrographs of *Botrytis cinerea* cells in chitosan-treated bell pepper fruit tissue. Labeling with the WGA/ovomucoid-gold complex. Fungal cells showed various degrees of alterations that range from cell wall loosening (**a**) and vacuolation (**a** and **c**) to retraction of degenerated protoplasm followed by deposition of material in the paramural spaces (**a** and **b**). The newly deposited material appeared preferentially labeled with gold particles (**c** and **d**). (Abbreviations are defined in Figure 1.)

membrane. The observed fungal alterations seem to be due to a direct effect of chitosan rather than to a process mediated via the host since similar alterations were also observed *in vitro* in the presence of chitosan. It is worthwhile to note that some of the alterations caused by chitosan were similar to those reported in aged and nutrient-deprived fungal cells.[67] This raises the question as to whether chitosan because of its polycationic nature and filmogenic property could have acted as a barrier to the outward flux of nutrients and consequently affect the establishment of a nutritional relationship between host and pathogen. This possibility merits further investigation and could add pertinent information concerning the mode of action of chitosan. The ultrastructural and cellular disorganization displayed by fungal cells in chitosan-treated tissue strongly suggests that the direct inhibitory effect of chitosan played a major role in the observed disease control.

C. INDUCTION OF DEFENSE REACTIONS

In addition to directly interfering with fungal growth, chitosan can induce a multitude of biological processes in vegetative tissue, including the stimulation of chitinase,[49] the accumulation of phytoalexins,[27] and an increased lignification.[50] In strawberry fruits, the ability of chitosan to stimulate defense enzymes such as chitinase seems to be expressed more in cut fruits than in intact ones.[20] This could make it an attractive preservative agent for cut and bruised fruits because of the interplay of its antifungal and eliciting properties. The eliciting ability of chitosan was also demonstrated with bell pepper and tomato fruits. In both crops, chitosan when applied as a stem scar treatment stimulated chitinase, chitosanase, and β-1,3-glucanase activity (Figure 3b and c). In chitosan-treated tissues, the activity of these enzymes remained elevated for up to 14 d after treatment. Because they can degrade fungal cell walls, these antifungal hydrolases are considered to play a major role in disease resistance.[23] The deliberate stimulation of lytic enzymes by chitosan could give the tissue a head start in restricting fungal colonization.

Ultrastructural studies revealed that chitosan treatment stimulates various structural defense responses in bell pepper and tomato fruits. Among the reactions observed, the most common ones were a thickening of host cell wall, frequent deposition of papilla, and occlusion of some intercellular spaces with fibrillar material (Figure 3a, e, and f). The various wall appositions were also found to be either completely or partially impregnated with amorphous electron-opaque substances (Figure 3a and f). This material is thought to contain antifungal phenolic-like compounds. While it is not possible to determine exactly the extent of the role played by host-mediated responses in the control of *B. cinerea*, the expression of inducible defensive reactions by chitosan seems to have contributed in restricting fungal infection. This is indirectly supported by: (1) the fact that invading hyphae were mainly restricted to the epidermal cells ruptured during wounding[64] and (2) the observation that fungal chitin was substantially reduced over fungal walls in contact with chitosan-treated tissue (Figure 1,e vs. f). Since chitin is covalently cross-linked with β-1,3-glucan in the wall of *B. cinerea*, the observed reduction in labeling is probably the result of synergetic action of chitinase and β-1,3-glucanase. Both enzymes were activated by chitosan treatment. The activation of systemic defense reactions that persist in the tissue by chitosan could also affect the resumption of quiescent infection. If this is the case, such treatment could be of significant importance since most postharvest diseases arise from latent infections that become active on the decline of the biosynthetic potential of the tissue to produce antimicrobial compounds.

D. PROSPECT FOR COMMERCIALIZATION OF CHITOSAN

Chitosan, a biodegradable food fiber, offers great potential as an antifungal preservative for fresh fruits and vegetables. Its potential value is attributable to the interplay of its

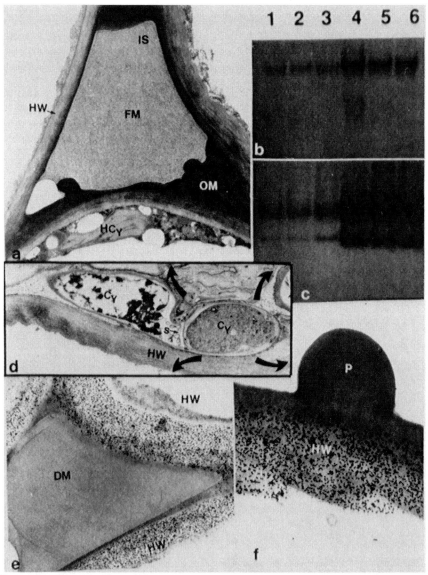

Figure 3 Transmission electron micrographs of chitosan treated bell pepper tissue and β-1,3-glucanase and chitinase activities after native gel electrophoresis (PAGE). Panels **b** and **c** show acidic β-1,3-glucanase and basic chitinase, respectively. Noninoculated control lanes (1 and 2) and chitosan-treated lanes (4 and 5) stored for 24 and 48 h. Inoculated control lane (3) and inoculated chitosan-treated lane (6) stored for 48 h. (**a, e**) Intercellular spaces filled with fibrillar material. Note the partial deposition of electron opaque material (**a**). (**d**) Severely altered fungal cell in contact with host wall. Note the retraction of degenerated protoplasm. Big arrows indicate various host defensive reactions that can further deter the establishment of an active lesion. (**f**) Papilla deposition along host cell wall. Note the regular and intense cellulose labeling over host walls. (Abbreviations are defined in Figure 1.)

antifungal and eliciting properties . The data accumulated so far bear witness to the film-forming ability of chitosan and considerably to its biological activities. Detailed studies on the permeability characteristics of chitosan films under high humidity to various gases and volatiles are required in order to establish the ultimate potential applications as a coating for harvested commodities. There are some indications in the literature on the unusual differential permeability of chitosan films to gases as opposed to typical plastic films, i.e., chitosan films are more permeable to oxygen than to carbon dioxide. Permeability of chitosan films may be modified either via chemical or physical changes. Above all, the antifungal and eliciting activities of chitosan present exciting possibilities for crop protection. Research must go on to reveal the mechanism by which this compound exerts its biological activities and to determine its compatibility with other available disease control methods, especially biological antagonists. Several antagonistic yeasts have been found to be compatible with chitosan.[64a] Combining antagonists with chitosan could represent a promising approach for the control of decay and ripening. Presently, chitosan is regarded as safe as indicated by feeding trials with domestic animals.[54] However, before it could be utilized as an antifungal preservative its safety for human consumption and its effect on the organoleptic quality of fruit need to be established. Furthermore, extensive postharvest storage tests are necessary to determine the feasibility of using chitosan coatings commercially.

VI. CONCLUDING REMARKS

Several promising biological control approaches that include use of antagonistic microorganisms, natural fungicides, and induced resistance are available for developing safer technologies for postharvest disease control. Among the proposed alternatives, development and enhancement of antagonistic microorganisms have been the most studied, and substantial progress has been made in this area. Attempts are now being made to market antagonistic microorganisms as a treatment for postharvest disease control. It is important, however, to realize that each of the advanced alternatives comes with limitations that can affect applicability. For instance, most antagonists have a limited spectrum of activity, and induced resistance is likely to provide only a short-lived protection since it relies on a protective potential that declines rapidly with ripening. As to plant-derived fungicide, there is a serious debate regarding their relative safety. More importantly, none of the available biological control methods have been shown clearly to offer consistent disease control comparable to that obtained with synthetic fungicides. This is further accentuated by the inherent variability associated with living systems.

In order for biological control to gain broader acceptance, its consistency and efficacy in controlling postharvest disease need to be enhanced to a level comparable to that of synthetic fungicides. This is unlikely to be accomplished through the use of the so-called "silver bullet" approach that utilizes a single biological control method. It has become apparent that a multifaceted approach that exploits the additive and synergistic effects of different treatments has advantages over the use of a solitary biological control method. For instance, the combination of UV-light treatment with antagonistic yeasts has been shown to further prolong the protection of fruits against postharvest disease incidence.[64b] Recently, chitosan has been found to be compatible with certain antagonistic yeasts.[64a] Combining chitosan with antagonists will make it possible to exploit the antifungal and eliciting property of chitosan as well as the biological activity of the antagonist. The complexity of the mode of action displayed by combined alternatives should make the development of pathogen resistance more difficult and present a highly complex disease deterrent barrier. Such multifaceted resistance strategies should also be expected to have greater stability and effectiveness than solitary methodologies.

REFERENCES

1. **Sommer, N. F.**, Strategies for control of postharvest diseases of selected commodities, in *Postharvest Technology of Horticultural Crops,* Kader, A. A., Kasmire, R. F., Mitchell, G. F., Reid, M. S., Sommer, N. F., and Thompson, J. F., Eds., Coop. Ext, University of California, Davis, 1985, 83.
2. **Eckert, J. W. and Raynayake, M.**, Host-pathogen interaction in postharvest diseases, in *Postharvest Physiology and Crop Preservation,* Liberman, M., Ed., Plenum Press, New York, 1983, 247.
3. **Eckert, J. W. and Ogawa, J. M.**, The chemical control of postharvest diseases: deciduous fruits, berries, vegetables and roots/tuber crops, *Annu. Rev. Phytopathol.,* 26,433, 1988.
4. **Wilson, C. L. and Wisniewski M. E.**, Biological control of postharvest diseases of fruits and vegetables: An emerging technology, *Annu. Rev. Phytopathol.,* 27, 425, 1989.
5. **Wisniewski, M. E. and Wilson, C. L.**, Biological control of postharvest diseases of fruits and vegetables: recent advances, *Hortscience,* 27, 94, 1992.
6. **Kuc, J.**, Plant immunization and its applicability of disease control, in *Innovative Approaches to Plant Disease Control,* Chet, I., Ed., John Wiley & Sons, New York, 1987, 255.
7. **Kuc, J.**, Immunization for the control of plant disease, in *Biological Control of Soilborne Plant Pathogens,* Hornby, D., Ed., C.A.B. International, Wallingford, U.K., 1990, 355.
8. **Collinge, D. B. and Slusarenko, A. J.**, Plant gene expression in response to pathogens, *Plant Mol. Biol.,* 9, 389, 1987.
9. **Lamb, C. J., Lawton, M. A., Dron, M., and Dixon, R. A.**, Signals and transduction mechanisms for activation of plant defense against microbial attack, *Cell,* 56, 215, 1989.
10. **Bowles, D. J.**, Defense-related proteins in higher plants, *Annu. Rev. Biochem.,* 59, 873, 1990.
11. **Dixon, R. A. and Harrison, M.**, Activation, structure, and organization of genes involved in microbial defense in plants, *Adv. Genet.,* 1990, 28, 165.
12. **Hammerschmidt, R. and Stermer, B. A.**, Induced systemic resistance to disease, in *Biochemical Plant Pathology,* Lamb, C. J., Dixon, R., and Kosuge, T., Eds., Elsevier, Amsterdam, 1984.
13. **Boller, T.**, Hydrolytic enzymes in plant disease resistance, in *Plant-Microbe Interactions,* Vol. 2, Kosunge, T. and Nester, E. W., Eds., Macmillan, New York, 1987, 385.
14. **Darvill, A. G. and Albersheim, P.**, Phytoalexin and their elicitors: a defense against microbial infection in plants, *Annu. Rev. Plant Physiol.,* 35, 243, 1984.
15. **Dewit, P. J. G. M.**, Induced resistance to fungal and bacterial diseases, in *Mechanisms of Resistance to Plant Diseases,* Fraser, R. S. S., Ed., Martinius Nijhoff, Dordrecht, 1985.
16. **Sequeira, L.**, Induced resistance: physiology and biochemistry, in *New Directions in Biological Control Alternatives for Suppressing Agricultural Pests and Diseases,* Alan R. Liss, New York, 1990, 663.
17. **Sequeira, L.**, Cross protection and induced resistance: their potential for plant disease control, *Trends Biotechnol.,* 2, 25, 1984.
18. **Caruso, F. L. and Kuc, J.**, Induced resistance of cucumber to anthracnose and angular leaf spot by *Pseudomonas lachrymans* and *Colletotrichum lagenarium, Physiol. Plant Pathol.,* 14, 191, 1979.
19. **Adikaram, N. K. B., Brown, A. E., and Swinburne, T. R.**, Phytoalexin induction as a factor in the protection of *Capsicum annum* L. fruits against infection by *Botrytis cinerea* Pers., *J. Phytopathol.,* 122, 267, 1988.

20. **El Ghaouth, A., Arul, J., Grenier, J., and Asselin, A.,** Antifungal activity of chitosan on two postharvest pathogens of strawberry fruits, *Phytopathology,* 82, 398, 1992.

21. **Mercier, J., Arul, J., Ponnampalam, R., and Boulet, M.,** Induction of 6-methoxymellein and resistance to storage pathogens in carrot slices by UV-C, *J. Phytopathol.,* 137, 44, 1993.

22. **Stevens, C., Lu, J. Y., Kahn, V. A., Wilson, C. L., Chalutz, E., Droby, S., Kawbe, M. K., Haung, Z., Adeyeye, O., and Liu, J.,** Ultraviolet light induced resistance against postharvest diseases in vegetables and fruits, in *Biological Control Postharvest Dis. Fruits Vegetables,* Proc. Workshop, Agricultural Research Service, U.S. Department of Agriculture, ARS-92, 1991, 160.

23. **Boller, T.,** Ethylene and the regulation of antifungal hydrolases in plants, *Oxf. Surv. Plant Mol. Cell Biol.,* 5, 145, 1988.

24. **Mauch, F., Mauch-Mani, B., and Boller, T.,** Antifungal hydrolases in pea tissue. II. Inhibition of fungal growth by combination of chitinase and β-1,3-glucanase, *Plant Physiol.,* 88, 936, 1988.

25. **Schlumbaum, A., Mauch, F., Vogeli, U., and Boller, T.,** Plant chitinases are potent inhibitors of fungal growth, *Nature, (London),* 324, 365, 1986.

26. **Keen, N. T. and Yoshikawa, M.,** β-1,3-endoglucanase from soybean release elicitor-active carbohydrates from fungus cell walls, *Plant Physiol.,* 71, 460, 1983.

27. **Kendra, F. D., Christian, D., and Hadwiger, L. A.,** Chitosan oligomers from Fusarium solani/pea interactions, chitinase/β-glucanase digestion of sporelings and from fungal wall chitin actively inhibit fungal growth and enhance disease resistance, *Physiol. Mol. Plant Pathol.,* 35, 215, 1989.

28. **Boller, T.,** Induction of hydrolases as a defense reaction against pathogens, in *Cellular and Molecular Biology of Plant Stress,* Key, J. L. and Kosuge, T., Eds., Alan R. Liss, New York, 1985, 247.

29. **Mauch, F. and Staehelin, L. A.,** Functional implications of the subcellular localization of ethylene-induced chitinase and β-1,3-glucanase in bean leaves, *Plant Cell,* 1, 447, 1989.

30. **Paxton, J. D.,** Phytoalexins — a working redefinition, *Phytopathol. Z.,* 101, 106, 1981.

31. **Bailey, J. A. and Mansfield, J. W.,** *Phytoalexins,* John Wiley & Sons, New York, 1982.

32. **Stoessel, A.,** Secondary plant metabolites in preinfectional and postinfectional resistance, in *The Dynamics of Host Defense,* Bailey. J. A. and Deverall, B. J., Eds., Academic Press, New York, 1983, 71.

33. **Haard, N. F. and Cody, M.,** Food and nutrition, in *Postharvest Biology and Biotechnology,* Dultin, H. O. and Milner, M., Eds., Wesport, U.S.A., 1978.

34. **Creasy, L. L. and Coffee, M.,** Phytoalexin production potential of grape berries, *J. Am. Soc. Hortic. Sci.,* 113, 230, 1988.

35. **Brown, A. E. and Swinburne, T. R.,** The resistance of immature banana fruits to anthracnose (Colletotrichum musae), *Phytopathol. Z.,* 99, 70, 1980.

36. **Haram, W.,** *Biological Effects of Ultraviolet Radiation,* Cambridge University Press, Cambridge, 1980.

37. **Bakken, A. K.,** Potential effects of ultraviolet B-radiation on plants and plant production in the north, *Nor. J. Agric. Sci.,* 3, 79, 1989.

38. **Chalmers, D. J. and Faragher, J. D.,** Regulation of anthocyanin synthesis in apple skin. I. Comparison of the effect of cycloheximide, ultraviolet light, wounding, and maturity, *Aust. J. Plant Physiol.,* 4, 111, 1977.

39. **Hahlbrock, K. and Scheel, D.,** Physiology and molecular biology of phenylpropanoid metabolism, *Annu. Rev. Plant Physiol. Plant Mol. Biol.,* 40, 347, 1989.

40. **Hadwiger, L. A. and Schwochau, M. E.,** Ultraviolet light-induced formation of pisatin and phenylalanine ammonia lyase, *Plant Physiol.,* 47, 588, 1971.
41. **Bridge, M. A. and Klarman, W. L.,** Soybean phytoalexin, hydroxyphaseollin, induced by ultraviolet irradiation, *Phytopathology,* 63, 606, 1972.
42. **Andebhran, T. and Wood, R. K. S.,** The effect of ultraviolet radiation on the reaction of *Phaseolus vulgaris* to species of *Colletotrichum, Physiol. Plant Pathol.,* 17, 105, 1980.
43. **Arul, J., Mercier, J., Baka, M., and Maharaj, R.,** Photochemical therapy in the preservation of fresh fruits and vegetables: disease resistance and delayed senescence, in *Proc. Int. Symp. Physiological Basis of Postharvest Technologies,* University of California, Davis, 1992, 42.
44. **Chalutz, E., Droby, S., Wilson, C.L. and Wisniewski, M. E.,** UV-induced resistance to postharvest diseases of citrus fruit, *J. Photochem. Photobiol.,* 15, 367, 1992.
45. **Ben-Yehoshua, S., Rodov, V., Kim, J. J., and Carmeli, S.,** Preformed and induced antifungal materials of citrus fruits in relation to the enhancement of decay resistance by heat and ultraviolet treatments., *J. Agric. Food Chem.,* 40, 1217, 1992.
45a. **Stevens, C.,** personal communication, 1993.
46. **Araki, U. and Ito, E.,** A pathway of chitosan formation in *Mucor rouxii* enzymatic deacetylation of chitin, *Eur. J. Biochem.,* 55, 71, 1975.
47. **Filar, L., Rahe, J. E., and Wirick, M. G.,** Bulk solution properties of chitosan, in *Proc. 1st Int. Conf. Chitin Chitosan,* Muzzarelli, R. A. A. and Pariser, E. R., Eds., MIT, Cambridge, MA, 1978, 169.
48. **Allan, C. R. and Hadwiger, L. A.,** The fungicidal effect of chitosan on fungi of varying cell wall composition, *Exp. Mycol.,* 3, 285, 1979.
49. **Mauch, F., Hadwiger, L. A., and Boller, T.,** Ethylene: symptom, not signal for the induction of chitinase and β-1,3-glucanase in pea pods by pathogens and elicitors, *Plant Physiol.,* 76, 607, 1984.
50. **Pearce, R. B. and Ride, J. P.,** Chitin and related compounds as elicitors of the lignification response in wounded wheat leaves, *Physiol. Plant Pathol.,* 20, 119, 1982.
51. **Averbach, B. L.,** Film-forming capability of chitosan, in *Proc. 1st Int. Conf. Chitin Chitosan,* Muzzareli, R. A. A. and Pariser, E. R., Eds., MIT, Cambridge, MA, 1978, 199.
52. **Elson, C. M., Hayes, E. R., and Lidster, P. D.,** Development of the differentially permeable fruits coating "Nutri-Save" for the modified atmosphere storage of fruit, in *Proc. 4th Natl. Conf. Controlled Atmosphere Research,* Blankership, S. M., Ed., North Carolina State University, Raleigh, 1985, 248.
53. **Bai, R. K., Huang, M. Y., and Jiang, Y. Y.,** Selective permeabilities of chitosan-acetic complex membrane and chitosan-polymer complex for oxygen and carbon dioxide, *Polym. Bull.,* 20, 83, 1988.
54. **Hirano, S., Itakura, C., Seino, H., Akiyama, Y., Nonaka, I., Kanbara, N., and Kawakami, T.,** Chitosan as an ingredient for domestic animal feeds, *J. Agric. Food Chem.,* 38, 1214, 1990.
55. **Hirano, S., Senda, H., Yamamoto, Y., and Watanabe, A.,** Several novel attempts for the use of the potential functions of chitin and chitosan, in *Chitin, Chitosan, and Related Enzymes,* Zikakis, J. P., Ed., Academic Press, Orlando, 1984, 77.
56. **Hadwiger, L. A., Fristensky, B., and Riggleman, R. C.,** Chitosan, a natural regulator in plant-fungal pathogen interactions, increases crop yield, in *Chitin, Chitosan, and Related Enzymes,* Zikakis, J. P., Ed., Academic Press, Orlando, 1984, 291.
57. **Benhamou, N.,** Ultrastructural and cytochemical aspects of chitosan on *Fusarium oxysporum* f. sp. *radicis-lycopersici,* agent of tomato crown and root rot, *Phytopathology,* 82, 1185, 1992.

58. **El Ghaouth, A., Arul, J., and Asselin, A.,** Potential use of chitosan in postharvest preservation of fruits and vegetables, in *Advances in Chitin and Chitosan,* Brines, J. B., Sandford, P. A., Zikakis, J.P., Eds., Elsevier Applied Science, London, 1992, 440.

59. **El Ghaouth, A., Ponnampalam, R., Castaigne, F. and Arul, J.,** Chitosan coating to extend the storage life of tomatoes, *HortScience,* 27, 1016, 1992.

60. **El Ghaouth, A., Arul, J., Ponnampalam, R., and Bollet, M.,** Use of chitosan coating to reduce water losses and maintain quality of cucumber and bell pepper fruits, *J. Food Process. Preserv.,* 15, 359, 1991.

61. **El Ghaouth, A., Arul, J., Ponnampalam, R., and Bollet, M.,** Chitosan coating effect on storability and quality of strawberries, *J. Food Sci.,* 56, 1618, 1991.

62. **Meheriuk, M. and Lau, O. L.,** Effect of two polymeric coatings on fruit quality of 'Barlett' and 'd' Ajou' pears, *J. Am. Soc. Hortic. Sci.,* 113, 222, 1988.

63. **El Ghaouth, A. and Arul, J.,** Potential use of chitosan in postharvest preservation of fresh fruits and vegetables, in *Proc. Int. Symp. Physiological Basis of Postharvest Technologies,* University of California, Davis, 1992, 50.

64. **El Ghaouth, A., Arul, J., Asselin, A., and Benhamou, N.,** Biochemical and cytochemical events associated with the interaction of chitosan and *Botrytis cinerea* in bell pepper fruit, submitted for publication.

64a. **El Ghaouth, A. and Wilson, C. L.,** unpublished data, 1993.

64b. **Chalutz, E. and Wilson, C. L.,** personal communication, 1992.

65. **El Ghaouth, A., Arul, J., Asselin, A., and Benhamou, N.,** Antifungal activity of chitosan on post-harvest pathogens: induction of morphological and cytological alterations in *Rhizopus stolonifer, Mycol. Res.,* 96, 769, 1992.

66. **El Ghaouth, A., Arul, J., Grenier, J., and Asselin, A.,** Effect of chitosan and other polyions on chitin deacetylase in *Rhizopus stolonifer, Exp. Mycol.,* 16, 173, 1992.

67. **Campbell, R.,** An electron microscope study of exogenously dormant spores, spore germination, hyphae and conidiophores of *Alternaria brassicola, New Phytologist,* 69, 287,1970.

INDEX